Downsizing Science

Downsizing Science
Will the United States Pay a Price?

Kenneth M. Brown

QI27
U6
B768
1998

The AEI Press

Publisher for the American Enterprise Institute
WASHINGTON, D.C.
1998

Available in the United States from the AEI Press, c/o Publisher Resources Inc., 1224 Heil Quaker Blvd., P.O. Box 7001, La Vergne, TN 37086-7001. To order: 1-800-269-6267. Distributed outside the United States by arrangement with Eurospan, 3 Henrietta Street, London WC2E 8LU England.

AEI wishes to thank the David and Lucile Packard Foundation for its support for this project.

Library of Congress Cataloging-in-Publication Data

Brown, Kenneth M., 1939-
    Downsizing science : will the United States pay a price? / Kenneth M. Brown.
      p. cm.
    Includes bibliographical references and index.
    ISBN 0-8447-4026-8. — ISBN 0-8447-4027-6 (pbk.)
    1. Research—Government policy—United States. 2. Federal aid to research—United States. I. Title.
IN PROCESS
338.973'06—dc21   96-32749
                                                                                                                                CIP
3 5 7 9 10 8 6 4 2

© 1998 by the American Enterprise Institute for Public Policy Research, Washington, D.C. All rights reserved. No part of this publication may be used or reproduced in any manner whatsoever without permission in writing from the American Enterprise Institute except in cases of brief quotations embodied in news articles, critical articles, or reviews. The views expressed in the publications of the American Enterprise Institute are those of the authors and do not necessarily reflect the views of the staff, advisory panels, officers, or trustees of AEI.

THE AEI PRESS
Publisher for the American Enterprise Institute
1150 17th Street, N.W., Washington, D.C. 20036

Printed in the United States of America

Contents

	FOREWORD, *Claude E. Barfield*	ix
	ACKNOWLEDGMENTS	xi
1	INTRODUCTION	1
2	THE END OF THE GOLDEN AGE Budget Pressures 5 An Overview of the Research Enterprise 9 The Age of Science 11 The Golden Age Comes and Goes 13 Conclusion 18	3
3	MEASURING THE COSTS OF DOWNSIZING R&D as a Source of Economic Growth 21 R&D as an Investment 25 Would Downsizing Be Costly? 28 Conclusion 33	20
4	A BROADER VIEW OF THE BENEFITS FROM SCIENCE More than Economic Benefits 35 Other Unique Aspects 41 Questionable Benefits 42 Conclusion 42	34

5	**A VITAL BUT LIMITED ROLE FOR GOVERNMENT** Reasonable Limits 46 Criteria for Evaluating Programs 48 Contrary Views 52 Conclusion 53	44
6	**JUST ANOTHER DOWNTURN?** Worrisome Signs 55 Attitudes toward Science 57 Ignorance of Science 58 Science and Higher Education 59 A Negative View of Scientists 60 Conclusion 61	55
7	**THE FEDERAL LABORATORIES—FIRST, DECIDE ON A MISSION** Downsizing Federal Laboratories 64 "Mission Lurch" 70 Living with Lower Funding 75 Conclusion 78	63
8	**THE ESSENTIAL FEDERAL ROLE IN ACADEMIC RESEARCH** The Washington Connection 82 Impacts of Downsizing 84 Conclusion 90	80
9	**IMPROVING THE ENVIRONMENT FOR INDUSTRIAL RESEARCH** What Price Federal Downsizing? 92 How Healthy Is Industrial R&D? 95 Reactions of Industry 101 Other Federal Policies and Funding 104 Conclusion 105	92
10	**INTERNATIONAL DIMENSIONS OF DOWNSIZING** U.S. Leadership, Past and Present 108 U.S. Leadership, Future 110 Looking Ahead a Few Years 114 Implications of Relative Decline 115 How Much Is World Leadership Worth? 117 That Old Shibboleth "Competitiveness" 120 Implications for Policy 121 Conclusion 124	108

CONTENTS

11	THE PRICE OF DOWNSIZING SCIENCE	125
	What Are the Alternatives? 128	
	Reflecting on the Future 132	

NOTES	133

INDEX	145

ABOUT THE AUTHOR	153

LIST OF FIGURES

2–1   Funding and Performance of U.S. Research and Development, 1996   10
2–2   U.S. Funds for Reseach and Development, 1957–1996   14
2–3   Science and Engineering Doctorates Awarded by U.S. Universities, 1957–1995   16
5–1   Private and Social Benefits from Research and Development   45
7–1   R&D Performed by the Federal Government, 1957–1996   65
8–1   R&D Performed at Universities, with Sources of Funds, 1957–1996   81
9–1   Federal and Industry Funds for Industrial R&D, 1957–1996   93
10–1   U.S. Share of World R&D Expenditures, 1961, 1993, and 2003   115

Foreword

KENNETH BROWN'S COMPREHENSIVE study of the potential impact of "downsizing science" is one of a number of studies the American Enterprise Institute has commissioned as a part of a major research project launched in 1996, entitled "The Science Enterprise in the United States." Other recent studies published in this series include an assessment of Vannevar Bush's *Science—The Endless Frontier* from the perspective of the 1990s, a multiauthored volume on future challenges to biomedical research, and an evaluation of the Clinton administration's Advance Technology Program.

In coming months, AEI will publish analyses of the difficult issues posed by intellectual property rights in the field of biotechnology and an evaluation of the economic results of public-private research and development partnerships.

In this study, the author reexamines the rationales that have been advanced for public support of scientific research and evaluates these rationales in light of current economic theory and political realities. He reviews and assesses the nature and scale of purported benefits to society from science. Using a "public goods" analytic framework, he then develops criteria for assessing federal programs and funding priorities.

In succeeding chapters, he applies those criteria to programs directed to the research universities, to the national laboratories, and to recommendations for public policy, including greater reliance on market solutions and the private sector. In a closing statement, he argues:

FOREWORD

Maintaining U.S. excellence in science is not an impossible dream, like ridding the world of crime or narcotics. It is not a problem of exploding costs, as with social security or Medicaid. It does not even require building vast new infrastructure and creating new institutional arrangements; most of what we already have is sufficient. The gap between excellence and mediocrity is measurable in the low billions of dollars per year. Clearly, continued excellence in science is within the nation's reach.

CLAUDE E. BARFIELD
Director, Science and Technology Policy Studies
American Enterprise Institute

Acknowledgments

I thank Neal Lane, Bennett Bertenthal, Cora Marrett, and Jeff Fenstermacher for granting me leave from the National Science Foundation to write this book. Equal thanks to Christopher DeMuth, Claude Barfield, and David Gerson for their generous hospitality at the American Enterprise Institute.

Thanks to these people who helped by reviewing drafts, providing information, and editing: Dennis Avery, Cynthia Beltz, William Blanpied, Jennifer Bond, Allyson Brown, Lisa Bustin, Eileen Collins, Ed Dale, Donald Dalton, Charles Dickens, James L. Edwards, Richard Florida, Murray Foss, Howard Gobstein, Mary Golladay, Mark Griffin, Margaret Grucza, Alan Hale, Eric Hanushek, Robert Helms, John Jankowski, Jean Johnson, Marvin Kosters, Kei Koizumi, Dana Lane, Charles Larson, Jennifer Lesiak, Clarisa Long, Ronald Meeks, Leonard Nakamura, Shanna Narath, Richard Nelson, Steve Nelson, Dan Newlon, Peter Sharfman, and Mark Symonds. Finally, thanks to my wife, Agnes, for her patience and support during the months this book was underway.

1
Introduction

THESE TRADITIONAL JUSTIFICATIONS for federal science funding are ingrained in discussions of science policy:
- Science produces tremendous benefits for the nation.
- The federal government must help support fundamental research because industry and private sector donors will not do enough.
- The United States must be a world leader, or close to it, in every field of science.

These principles, however, are inconsistent with current realities. Federal funding of science has been reduced and will be reduced more, regardless, apparently, of whatever benefits may be lost. Some have proposed more direct federal assistance for industry, even though industrial research is growing. Scientific advances in other countries, especially in Asia, may erode the U.S. position of world leadership. Yet current plans for downsizing science endanger fundamental research in particular, where the rationale for federal support is strongest and which is vital to continued world leadership.

What is going on here? Are the principles wrong, or the policies, or both? One thing is clear: federal funding for all "discretionary" programs, including science, is under extreme pressure that will surely increase during the next several years and into the next century. Significant reductions in federal funding for scientific research seem inevitable.

This book reexamines the rationales for the governmental role in science in an effort to shape an efficient policy response to tighter federal budgets. First, it looks at the nature and magnitude of the benefits from science. Next, it draws on the "public goods" argument for federal support to develop criteria by which to evaluate federal science funding: which programs should have the highest priority for continued support and which could be dispensed with most easily? How to define and allocate intellectual property rights clearly looms large in this new context: with greater dependence on nonfederal funds, the nation's legal framework for intellectual property rights will become increasingly important.

Finally, the book addresses how the United States will fare with a downsized science establishment: the fact of downsizing is at odds with the often-stated goal of world leadership in science, particularly when certain other nations are expanding their support for science.

The nation's research effort is in several parts—federally performed research, federally financed research outside the government, and industrial research. All will be affected by the coming changes. This book recommends specific actions, including in some cases greater reliance on markets and the private sector, to obtain the best results from the diminished federal support of science.

2
The End of the Golden Age

DESPITE EXTENSIVE EVIDENCE that scientific research brings vast social and economic benefits, the federal budget for research and development (R&D) is being reduced, possibly very considerably. Why is this happening—and what will be the consequences? First, a historical perspective might be useful in understanding scientific research and its funding.

As the millennium nears, the industrial countries of the world have achieved living standards unprecedented in human history. Much of this progress is the result of scientific research. Without the previous centuries of research and its application through new technologies, long-term economic growth would have been much slower, more along the gloomy lines that Malthus predicted.[1] During the past several decades, science and technology have been especially productive, bringing longer life spans, better medical care, faster communications, higher labor productivity, new products resulting from research in electronics, biotechnology, and materials science, and generally higher living standards than have ever been known.

This process needs to continue. Many scientific and technological challenges remain, involving health, space, the environment, energy, national security, and the understanding of our natural and social systems. Scientific research can contribute to meeting each of these challenges. The words Vannevar Bush used to begin *The Endless Frontier* remain relevant:

> Progress in the war against disease depends upon a flow of new scientific knowledge. New products, new industries, and more jobs require continuous additions to knowledge of the laws of nature, and the application of that knowledge to practical purposes. Similarly, our defense against aggression demands new knowledge so that we can develop new and improved weapons. This essential, new knowledge can be obtained only through basic scientific research.[2]

Nonetheless, there is strong evidence that the golden age of U.S. science is over, that there is a decline—or at best a leveling out—in practically every statistic that describes scientific activity. Most conspicuous is the expected downsizing of federal funding for R&D.

It is easy to lose sight of just how recently the federal government entered the world of science and how profoundly it has altered the scope of the scientific enterprise during the past sixty years. Nor is it widely recognized that the federal role in science may be changing very substantially, just as it has in other realms of governmental activity. Federal support of R&D, once on a rapid growth path, has begun to decrease and is likely to continue doing so. This decline will probably have serious repercussions, not just for the scientists that perform the research and the federal, industrial, and university laboratories where they work, but potentially for the U.S. economy, which has come to rely heavily on a steady stream of scientific "output."

Financial support for science is not anywhere near the top of the political agenda, except within the scientific community. One would be hard pressed to find more than a passing mention of science policy in all the speeches, debates, talk shows, and sound bites of the 1996 political campaigns. Indeed, complacency about science appears almost universal, apparently based on the belief that continued scientific progress can be taken for granted regardless of how federal policy is shaped. Edison didn't get any federal support, did he?

Just how serious a problem is this prospective downsizing of science? Will there be an economic price to pay? Are the benefits of science sufficiently understood and clearly linked to federal support? Or is the science community overstating its own importance in terms of actual benefits delivered to society? How can the scientific enterprise accommodate itself to tighter budgets? It is these and related questions with which this book will deal.

Budget Pressures

The two seven-year balanced-budget plans presented by Congress and the Clinton administration early in 1996 shocked the scientific community. Both plans called for cuts in civilian research and development on the order of 20 to 25 percent over the seven years ending with 2002.[3] Defense R&D, already shrinking, appeared to be in for larger reductions. Even more traumatic, these proposed cuts were to be imposed on an R&D budget that had already been declining in real terms for several years. According to National Science Foundation (NSF) data, federal funds for R&D, adjusted for inflation, actually peaked in 1987 and have drifted down slowly since then.[4]

Of course, budget plans such as these are notoriously unstable: none of the many budget plans that stemmed from the Gramm-Rudman Act in the 1980s, for example, has been fulfilled. Thus, studying the latest balanced-budget plans closely is a waste of time, because they will certainly change in ways we cannot predict. According to the latest figures from the administration in the proposed FY 1998 budget, federal support for R&D would decline by 16.8 percent in real terms between FY 1994 and FY 2002. Within this total is a 13.7 percent cut for nondefense R&D. These smaller reductions, however, reflect mainly changes in the economic assumptions in the budget, not greater generosity toward science.[5]

The point here is that large cuts were seriously proposed by a Democratic president and a Republican Congress, the highest elected officials of government who were thought to be, and considered themselves to be, friends and supporters of science. The budget plans showed that science is not a protected public investment but a target for reduction, on a par with other federal programs.

Moreover, the 1996 balanced-budget plans were hardly spur-of-the-moment political gambits. They reflected powerful political and economic forces that are likely to continue and possibly intensify for some years to come. Indeed, both political parties recognize those forces and have responded with rhetoric and at least some action. President Clinton has said, "The era of big government is over." The GOP congressional victory in 1994 had at its heart a deep aversion to the big-government programs that had built up over the previous decades. Efforts to pass a balanced-budget constitutional amendment represent another approach to relieving the long-term stress in the federal system.

Longer-Term Budget Pressures. The federal budget deficit has been abnormally high for peacetime since the early 1980s. The deficit declined substantially in the mid-1990s, partly because of the robust economy, but the remaining savings to eliminate the deficits altogether will be more difficult to achieve than the earlier ones. The deficit could easily balloon again with a recession, with excessive tax cuts, with unforeseen financial or national security disasters, or with faltering congressional resolve to reduce spending in the big entitlement programs. It is entirely possible that Congress will enact new spending measures and tax breaks, the nature of which cannot now be foreseen, thereby putting more pressure on science.

At the heart of the problem lie the entitlement programs, where spending grows automatically with caseloads and other factors and is not controlled by annual appropriations. Without explicit actions to change them, entitlement programs will increase substantially during the next several years. It is estimated, for example, that Medicare's hospital insurance trust fund will be depleted in 2001, with shortfalls in net revenues before that time measured in the hundreds of billions of dollars. And in 2001 the oldest baby boomers will still be only fifty-five years old, years short of mass retirement.

The outlook deteriorates dramatically after 2010, when the huge baby-boom generation will begin to draw benefits from the three largest entitlement programs—social security, Medicare, and Medicaid. At the same time, growth in revenues will slow because the proportion of people working and paying taxes will shrink. Over the next thirty-five years, the Social Security Administration estimates that the number of people aged sixty-five and older will double, while the number aged twenty to sixty-four will increase only 20 percent. In 1960, there were about twenty social security beneficiaries for every 100 workers; by 2030, it will be fifty beneficiaries per 100 workers, and they are slated to receive a benefits package that is far more generous than it was in 1960. As a result, the deficit will increase rapidly after 2010. It is hard to envision any plan to deal with this spending explosion that does not come down hard on discretionary spending.[6]

The so-called peace dividend, once touted as the solution to budget problems, is seldom heard of any more. Since the dissolution of the Soviet empire in 1989, defense spending has fallen by hundreds of billions of dollars below the amounts projected in the last of the cold war budgets. There was indeed a peace dividend, but it has disappeared into the huge maw of entitlements. Diminished military procurement helped accomplish the spending reduction, but knowledgeable observers now believe that the reserve of military materiel is now much diminished and that procurement will have to rise sharply, this aside from any new national security threats or expanded military missions that may arise in the coming decade.[7] Interest on the debt, now 15 percent of the federal budget, will continue at a high level, although its share of the budget will not change much if deficits are reduced according to plan.

Even in this discouraging context, pessimism about the future of science should not be overstated. After the current and prospective reductions in federal R&D spending, the federal science effort is and will continue to be very large, particularly when measured against anything but the recent past. We are not facing a reversion to the world of sixty years ago, when the federal government was essentially a nonplayer in research and development. We are, though, facing a world in which choices will be more difficult and some benefits of science are likely to be forgone.

The FY 1998 Budget. With "mandatory" outlays, chiefly entitlements, large and growing, domestic "discretionary" spending unavoidably comes under heavy pressure. That category, which is controlled by annual appropriations, is where all the R&D spending is to be found. Domestic discretionary spending has already declined to only 16 percent of the total budget. The budget resolution for the 1998–2002 period passed by Congress in 1997 calls for virtually no growth in the dollar total for this category over the five-year period and a decline of $115 billion in real terms.

Over the past several decades (the Apollo program excepted), nondefense R&D spending has closely tracked with nondefense discretionary spending. Total federal R&D, including defense, has hovered at about 14 percent of discretionary spending, year after year. Moreover, discretionary spending is now capped by law as part of the five-year plan to balance the budget, which puts greater limits than ever before on the ability of each Congress to increase this type of spending. As a result, if discretionary spending continues to be targeted in the quest for a balanced budget, as appears very likely, then further cuts to federal support for research are inevitable.[8] Given the strong and continuing political demands for increases in other parts of the discretionary budget and for tax cuts, R&D will be hard pressed to hold onto its traditional share of discretionary spending.

The Clinton administration's handling of the science budget offers some clues to how science and technology will fare in the effort to balance the budget. For FY 1998, the president's budget for R&D calls for a slight increase (0.6 percent over FY 1997) in current dollars but a decrease of 1.9 percent in constant dollars,[9] to just over $75 billion.[10] Within this total, defense R&D declines by 4.0 percent, and nondefense R&D increases by 0.6 percent (both in constant dollars). As for changes from FY 1994, only two agencies are slated for constant-dollar increases: the National Institutes of Health and the National Science Foundation. Basic research, particularly on health, comes out the best, defense comes out the worst, and technology is somewhere in the middle.

The Shift to the Private Sector. Some argue (or hope or assume) that private companies will pick up the slack left by federal downsizing.[11] Yes, industrial R&D is relatively healthy, and companies that find it profitable will fund a great deal of research. But while private firms would probably do somewhat more if the government draws back, they would not come anywhere close to filling a gap of the magnitude we have suggested might occur.

Furthermore, companies would not fund the same type of research that the government abandons. Firms concentrate on their own product development and on research that they can "own," in contrast to so much of the federal research: fundamental research—with wide application and limited appropriability—and research directed to national needs, such as defense. To the extent that industry provides more funding for university-based research, it will probably pull researchers away from fundamental research toward the specific goals of whichever company pays the bill.

Nevertheless, the data clearly show that over the past twenty years or more the private sector has been funding a larger share of the nation's R&D. We need to understand why this is happening and what lessons it holds for making federal downsizing more efficient. We need to determine how the private sector could be induced to do more research, with the right market incentives and with fewer impediments. Chapters 7, 8, and 9, which look at federally performed science, federally supported university research, and industrial research, discuss various modes of privatization.

An Overview of the Research Enterprise

Let us take a closer look at the R&D enterprise: the institutions, the people, and the money. First, some definitions. *Research* is a systematic study directed toward more complete scientific knowledge or understanding of a subject. Federal data classify research as either basic or applied according to the objective:

- The objective of *basic research* is to gain knowledge or understanding of phenomena without specific applications in mind. Its results are generally uncertain, both when it begins and even after it is over because they are long term and diffuse. "Fundamental" research means the same thing.
- The objective of *applied research* is to gain knowledge or understanding necessary for meeting a specific need, rather than to expand the frontiers of knowledge. Industrial research is more typically applied than basic. In this book, *science* is often used interchangeably with *research*.
- *Development* is the systematic use of the knowledge or understanding gained from the research to solve the technical problems involved in bringing a new product or process into production. It excludes quality control, routine product testing, and production. Engineering activities predominate in "development," although it should be recognized that the term *research* also covers research in engineering.

This basic-applied-development breakdown is not clear-cut, as the three categories blend into each other at the borders between them. Some writers, like Donald Stokes, have proposed other intellectually appealing taxonomies,[12] but they do not match the existing data system. A more practical modification of the R&D taxonomy was proposed in a 1995 National Academy of Sciences report.[13] It recognized that R&D, at the development end of the spectrum, blends into testing, evaluation, and other activities that do not belong in R&D at all. Without these activities, mainly in the Defense Department's budget, what the report defines as the "federal science and technology" (FS&T) budget is around $35 billion or $40 billion, around half of what we usually call feder-

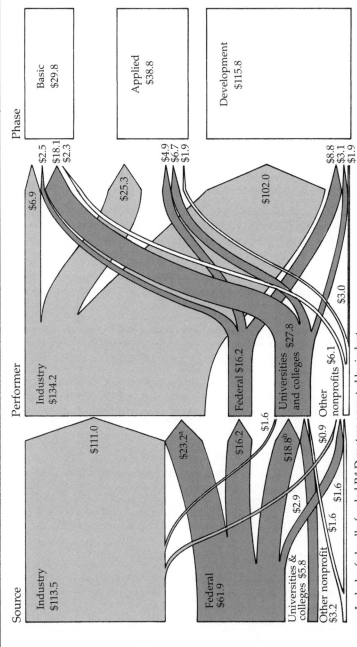

FIGURE 2-1
FUNDING AND PERFORMANCE OF U.S. RESEARCH AND DEVELOPMENT, 1996
(billions of dollars)

a. Includes federally funded R&D centers operated by industry.
b. Includes federally funded R&D centers operated by universities.
SOURCE: National Science Foundation.

ally funded R&D. Basic and applied science and technology are treated as one interrelated enterprise. This, in my view, is a constructive change in the way we look at R&D budgets, but it has not found general acceptance.

Basic research does *not* mean research that is divorced from all practical ends.[14] In fact, much of basic research is targeted to areas of science that have well-defined payoffs, as in health and agriculture. The "basicness" of the research is in its goal of understanding the underlying, or basic, processes of nature that will eventually enable applications of that understanding to specific technical problems.

The main actors in the scientific enterprise are the federal government, universities, companies, and nonprofit institutions. Each of these funds research and performs research. Figure 2–1 illustrates how the money flows from funders to performers and the split between fundamental and applied research. It shows that the government is mainly a funder of others, although it performs a considerable amount of research. Industry spends most of the money, performs most of the R&D, and focuses on development. The bulk of basic research is performed at universities, which depend heavily on federal funding.

The Age of Science

The most important contributions of science to the world are not very old. If we attempted to name the 100 greatest scientists in history, all but a very few of them would be from after 1800, and the great majority would be from this century. Einstein, Bohr, Fermi, and the other pioneers of nuclear physics were of the twentieth century. Watson, Crick, Feynman, Bardeen, Hubble, Pauling, von Neuman—all were fairly recent.[15]

Large-scale federal support for R&D is of even more recent vintage. In fact, many of the scientific greats performed their work well before the era of massive federal support. Some of the earlier scientists got a little support from the government, but this support went just to the select few, not to hundreds of thousands of researchers as occurs today. Corporate research was similarly sparse and fragmented during these bygone eras. Thus, while it is natural to see this decade's cuts in federal R&D spending as a reversal of a trend, over the longer span of history the three postwar decades may turn out to be an anomaly.

Early federal support for science was scant, sporadic, and oriented toward practical technologies. In 1836, for example, Congress spent $30,000 to help Samuel Morse construct an experimental telegraph line between Washington and Baltimore. During the nineteenth century, agriculture was the main focus of federally supported research. That support took root in 1862, when the newly established Department of Agriculture planted a garden on the Mall (about where the Air and Space Museum now stands) and began publishing research bulletins. By 1888 experimental stations had been set up in all thirty-eight states.[16] The land-grant colleges established by the Morrill Act in 1862 gave another push to agricultural research. Over the following century, technical progress was rapid in agriculture, with production and productivity growing while the farm labor force shrank. Given this history, it is understandable that agriculture became the first sector in which economists subjected the economic impact of R&D to rigorous scrutiny.[17] The success of this applied form of research lent support to proponents of federal support in other fields.

In the early decades of the twentieth century, the federal role grew to include research on public health, national security during World War I, and small efforts to help U.S. business, mainly in aeronautics and standards. As late as the 1930s, federal funding for science was small and scattered among isolated programs. One estimate puts total R&D at a mere 0.6 percent of GNP in 1934,[18] which was about the norm for industrial nations at the time. Support for academic research in 1935 from all sources was approximately $50 million;[19] adjusting for inflation (which is inexact over such a long span of years), academic research spending has increased since then by a factor of around fifty or sixty. Thus, before World War II, U.S. spending on science was minimal by today's standards and unexceptional by contemporary international standards.

During World War II, Vannevar Bush, an MIT engineer, became head of the White House Office of Scientific Research and Development, which mobilized scientists for war-related research efforts such as the Manhattan Project that developed the atomic bomb. President Roosevelt asked Bush for a report on how the federal government could promote scientific progress in the postwar era. The report, *Science—The Endless Frontier*, given to President Truman in 1945, became the single most influential document shaping the nation's R&D policy since that time.

While the Bush report made many specific recommendations, it is best known for its powerfully stated case that a steady flow of new scientific knowledge was essential to social and economic progress. The argument was readily accepted at high levels of government, particularly in light of the spectacular military results of research, which, Bush believed, could be turned to many peacetime goals. Bush emphasized the importance of fundamental research, which he believed would help applied research in industry deliver the actual benefits. In the postwar years, the federal government became the principal supporter of university research, and during these years the university system grew rapidly and increased its dependence on government funds for research.

The Golden Age Comes and Goes

The current outlook for science is all the more unsettling because it comes after rapid expansion (figure 2–2). Between 1953 and 1970, total spending—public and private—on research and development grew at an average annual rate of 6.7 percent, adjusted for inflation. This figure was more than double the growth rate of the economy. Although Vannevar Bush's recommended department of science never came into being, this massive spending on scientific research certainly fulfilled (or perhaps greatly exceeded) his main objective.

During the 1970s, R&D spending slowed, owing in large part to the phasing down of the space program, and annual growth, adjusted for inflation, averaged just 1.6 percent. During the 1980s, the growth of R&D spending rose to 4.3 percent per year. Over the whole period, 1953 to 1990, average annual growth was 4.8 percent, about two percentage points above the growth of the economy as measured by gross domestic product.

As the 1990s began, however, the picture changed. Between 1990 and 1996, total R&D spending grew at a rate of only 0.8 percent, about two percentage points slower than gross domestic product. During that time, federal funding fell in real terms by an average of 2.3 percent per year, so that by 1995 federal funding of R&D was more than 15 percent below its peak year of 1987.[20] Thus, according to this book's basic premise, if the budget plans are

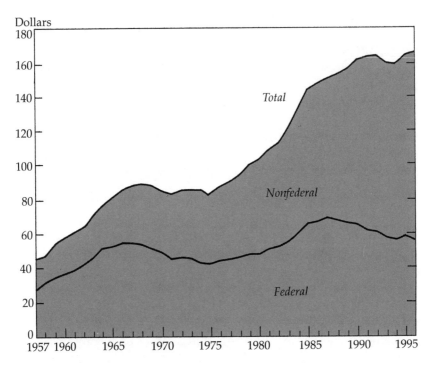

FIGURE 2-2
U.S. Funds for Research and Development, 1957–1996
(billions of 1992 dollars)

Source: National Science Foundation.

carried out, federal funding of R&D will fall by 16.8 percent between 1994 and 2002, with adjustment for inflation.[21]

Company funding of R&D in real terms was about level during the 1990s; some of the growth was in part the artifact of a change in the way the numbers were collected.[22] It appears that company spending is recovering from a slack period and is back on its long-term growth path.

Downsizing has international dimensions that need to be viewed in the context of other nations' research spending. The U.S. share of the industrial world's R&D spending fell steadily from its peak right after World War II, to around 71 percent in 1961, to about 43 percent in the middle 1990s. The projected budget cuts, together with likely R&D spending increases elsewhere, particularly in Asia, would take the U.S. share below 30 percent by 2002.[23]

Thus, even though Washington is unlikely to stick precisely to its long-term budget plans, and discretionary caps may be raised a little as the pressures to increase spending prove unbearable, it seems clear that federal science spending will continue to decline in real terms. The nation has already entered a period of austerity for R&D funding, starting during the late 1980s.

Some, like Cal Tech's David Goodstein,[24] would argue that the golden age of science was over even earlier than that; this interpretation of the data is also plausible. The space program was a huge bump in federal funding for R&D that grew rapidly in the 1960s and shrank in the 1970s. The defense buildup that started under President Carter and reached its height during the Reagan administration was another large upsurge that proved temporary. If one dates the beginning of the decline from the end of the Apollo program and treats the later defense buildup as an anomaly (and also because it dealt so heavily with development as opposed to basic and applied research), then the beginning of the decline can be dated back in the 1970s, as Goodstein suggests.

Tracking the number of Ph.D. recipients gives another perspective on the rapid rise of science (figure 2–3). The number of Ph.D.s awarded in science and engineering exploded in the post-Sputnik years, growing by 9.4 percent annually from 1957 through 1971. After a lull in the 1970s, Ph.D. awards were on the rise again, and many would characterize the current market as one of oversupply of new scientists and engineers. Unemployment rates, while low in comparison with those in the rest of the work force, are now higher than they were in the late 1980s. New Ph.D.s take longer to find jobs, and it has become much more difficult to land tenure-track positions at research universities.[25]

So in sum, what we have observed is a vast upsurge in science that began in the 1940s, a continuation and expansion of defense R&D after the war, the strong growth of company-funded research, an expensive space race, a health research bonanza, and a gradual spread of funding to nonmilitary research in many different fields, including the environmental and social sciences. In this perspective, federally supported R&D is a relatively young phenomenon. Indeed, taking a longer-term perspective, the golden age of science may be just an anomaly, a blip on history's radar

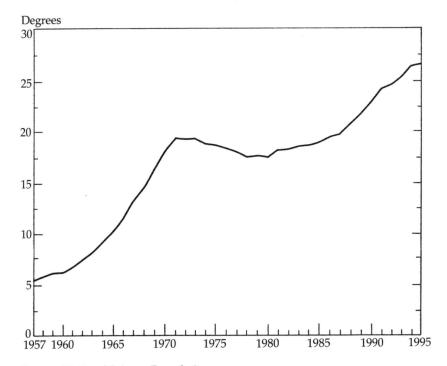

FIGURE 2–3
SCIENCE AND ENGINEERING DOCTORATES AWARDED BY U.S. UNIVERSITIES, 1957–1995
(thousands)

SOURCE: National Science Foundation.

screen. Yet, federal support has become well entrenched in the system and cannot be withdrawn without considerable dislocations.

Cyclicality in Federal Funding. This brief chronology may give the misleading impression that the recent history of science funding is simply one of postwar boom and recent bust. More detailed examination of the numbers shows several peaks and troughs over the years, particularly in the individual budgets of the funding agencies. NASA has experienced not just one peak in funding at the time of the Apollo mission but another related to the space shuttle, and possibly another with the space station. Defense research has had its ups (Vietnam War, Reagan administration) and downs (currently). Basic research has ebbed and flowed. Fluctuations in industrial research result largely from instability in that part of the funding that comes from the government. Health research is now booming, but funding may have started to level out.

This cyclicality causes problems of its own. In economics, the "accelerator" effect refers to the observation that changes in the demand for a good can cause a much larger change in the demand for new capital stock used to produce the good. For example, during the depression of the 1930s, the amount of capital goods required decreased, causing net investment actually to go negative for a year or two. Similarly, in science a slowdown in funding for research brought about an abrupt drop in the demand for new scientists to perform that research. This effect will be seen a number of times in later chapters as we consider the effects of downsizing on the producers of research.

The past record of cyclicality in federal funding gives rise to the idea that the current downsizing may be just another downturn from which science will recover. Above, we have presented evidence that the budget crunch is likely to intensify for some years. In chapter 6, we examine the possibility that additional factors may make this spending downturn last for a long time.

Why Downsize Now? Why this most recent reversal, this retreat from the postwar consensus that scientific research on a grand scale was a good investment of public funds? If this policy were no more than a reaction to the newly popular wish to reduce the federal deficit, then we could hope that a straightforward exposition of the benefits of science would be enough to build a fence around the science budget.

But science funding faces other problems. With the collapse of the Soviet Union, the defense rationale for federal support lost much of its power.[26] When national survival is at stake, the political imperative is clear: spend whatever it takes to become as safe and secure as science can make us. Now that our survival is no longer in doubt and none of our prospective opponents are a match for the armed forces we already have, then we have a plausible excuse for easing support for both military and civilian research.

Moreover, unsavory charges regarding academic fraud and quarrels between federal administrators and universities over the costs of research have weakened political support for public funding of science. Finally, some are fearful that the science enterprise has become too vast and complex, a huge labyrinth compared with the clear vision of Vannevar Bush, and that we should therefore carefully review that enterprise to determine which parts of it remain essential; recall President Eisenhower's warning in his farewell address. While his words of caution about the "military industrial complex" are the best remembered part of his speech, he also worried about achieving the right balance in science and technology between the government and the scientists it funds:

> In [the technological] revolution, research has become central; it also becomes more formalized, complex, and costly. A steadily increasing share is conducted for, by, or at the direction of, the Federal government . . . . The prospect of domination of the nation's scholars by Federal employment, project allocations, and the power of money is ever present—and is gravely to be regarded. Yet, in holding scientific research and discovery in respect, as we should, we must also be alert to the equal and opposite danger that public policy could itself become the captive of a scientific-technological elite.[27]

Conclusion

This chapter has described science largely in quantitative terms: how much are we spending on science, and by how much might this be reduced? But the quantitative analysis is only a piece of the story. Scientific research is, after all, a unique part of the economy. Researchers in fundamental science, by and large, are paid to do pretty much what they want to do, in hopes that something worthwhile will come out of it. Millions of people engage in R&D, with thousands of others staffing the complex operation of getting funds to what they hope are the best research projects. The efficiency with which the institution operates is just as important as the amount of money flowing into it.

Every measure of the scientific enterprise given in this chapter was a measure of inputs: funding, number of scientists, and so forth. But the main concern of public policy is outputs—what does the nation get for its money, and what will happen if we cut funding by 10 or 20 percent? Outputs of science are uncertain and diffuse, with a distant payoff. Outputs are hard to measure after the fact, let alone in advance of the research activity. It is very hard to buy research results; we can buy only inputs (scientists, laboratories), allocate these inputs according to broadly defined priorities, and hope for the best.

The next chapter will begin to deal with the benefits of research, giving rough measures of the impact that prospective budget cuts might have. Later chapters discuss other aspects of the benefits of research and address the complexities of how the research enterprise is likely to react to changing financial signals.

3
Measuring the Costs of Downsizing

WHAT DO WE KNOW ABOUT the magnitude of the benefits that come from scientific research? What price would the nation pay for downsizing science?

The accomplishments of scientific research have so thoroughly permeated modern life, and have so many dimensions, that they defy precise measurement. Individual projects and discoveries interact in unpredictable and complex ways. Today's scientific discoveries build on the knowledge accumulated from the past, a process that continually upgrades the value of past research. Even the term *spillover effects*, meaning the benefits that flow to those who had nothing to do with performing the research in question, connotes something indistinct or nebulous.

Technological progress is of such obvious importance to economic growth that economists have devoted tremendous efforts to measuring its effects. Their results are a good starting point. First, we will work through some rough estimates based on the measurements of the economic effects of technological change. We will point out the limitations of the estimates as they pertain to the costs of downsizing, but we will not go deeply into the theoretical and statistical problems that afflict these studies.[1]

The general model we are using here is this: research produces new knowledge, which leads to technological change, which results in more output per worker.[2] This progression is inexact, particularly the link between new knowledge and technical change. Some research leads to nothing, and some technological change takes place apparently independently of any particular prior research.

The last part of the chain of causation is very strong, however. It is absolutely established that technological change is a major source of economic growth. In fact, it is practically a tautology if we consider technological change to be "everything that causes increases in output besides increases in inputs of capital, labor, and materials." Technical change, defined this broadly, is the most important source of growth, accounting for between 80 and 90 percent of growth in output per worker during the course of the twentieth century.

Measuring the entire chain of causation, what economic benefits result from a given amount of research, is harder. To the question of how much scientific research has contributed to economic growth per capita and to our standards of living, the capsule response of the body of economic literature can be fairly stated as, "We're confident that the contributions of science have been large, and we can give you some ballpark numbers, but you must bear in mind all sorts of qualifications and caveats."

These caveats are particularly important when research results are used as they are in this chapter, to estimate economic losses that would result from reducing science funding. We will detour around a swamp of technical qualifications, giving the rough results first and then discussing their relationship to reality. The object of this discussion is to give a general idea of the magnitude of the economic benefits of science, not to expose the technical problems with this or that estimate. We must also recognize that this chapter takes a fairly limited view of the benefits of science, considering only the most easily measurable economic benefits. Chapter 4 takes a more expansive approach to the results of scientific research.

R&D as a Source of Economic Growth

Economic writers have long appreciated the central role of technological change, but only in recent decades have any useful quantitative results emerged. Robert Solow, in his famous 1957 article, helped pave the way for one approach.[3] He noted that between 1909 and 1949 U.S. economic output had about doubled, an increase far in excess of the growth in inputs of labor and capital. Put another way, the growth of output per worker was considerably more than could be ascribed to growth in the amount of capital—machines, land, factories—per worker. Solow concluded that technological change, defined very broadly, accounted for seven-eighths of the growth in productivity during the period he examined.

Now we come to the much more difficult part of the model: how much does scientific research contribute to technological progress and hence to economic growth?

More recent studies have updated, refined, and expanded the Solow results to address this question. Edward Denison, for example, statistically separated the sources of growth into various components and concluded that technological change, or what he called the "advance of knowledge," accounted for about 40 percent of the total increase in U.S. national income per person employed during 1929–1957.[4] In later work he estimated that the advance of knowledge accounted for similarly large contributions to the growth of output per unit of input for other periods of time.[5]

The Labor Department's Bureau of Labor Statistics (BLS) periodically estimates the contributions of the components of the growth of labor productivity. In a 1996 release,[6] BLS reported that research and development contributed about 0.2 percentage point per year to productivity growth between 1948 and 1994, when the average annual rate of growth in output per hour was 2.4 percent. In a more recent period, 1979 to 1994, the contribution of R&D was also 0.2 percentage point, although productivity growth had slipped to about 1.2 percent. (Additions to the capital stock and changes in the labor force are the other broadly defined components of total productivity growth. Note too that the BLS causal factor, R&D, is much narrower than Denison's "advance of knowledge," and so it is to be expected that, according to the BLS, R&D accounts for a smaller portion of growth than Denison's formula does.) Thus, in the more recent period, R&D contributed about one-sixth of the growth in productivity.

All these estimates are fairly rough. Skeptics like to quote economist Moses Abramowitz, who said that the Solow-type residual "is a measure of our ignorance." But Abramowitz made his comment forty years ago, when this branch of research was in its infancy. Substantial advances in understanding have occurred since that time.[7] Data limitations still prevent a full comprehension of the phenomena of technical change, and they will probably always prevent exact measurement of all the effects. But such limitations should not lead us into the non sequitur "if we can't measure an effect, then it doesn't exist." The fault may lie in our measuring tools.

If we grant that R&D has been a major contributor to the productivity of the economy, how can we quantify its importance? One approach is to assume that no R&D had been performed between, say, 1963 and 1992 and calculate the resultant change in gross domestic product (GDP). Using the BLS estimates cited above, I calculated how much lower GDP would have been without the productivity increment contributed by R&D. The growth of GDP was reduced by only a little at first but by more and more as the difference between actual and hypothetical R&D compounded. By my calculations, 1992 GDP would have been 5.5 percent less than it actually was, or around $330 billion in that one year alone.

These figures, of course, reflect the wholly implausible scenario of reducing research to zero. They also assume that the alternative use of the R&D funds would not contribute to GDP growth. Moreover, it is doubtful that the BLS figures can be used legitimately to estimate such an enormous change in R&D that never came close to happening during the years for which the estimates were made. More reasonably, the analysis of downsizing should be confined to smaller, more realistic changes.[8] Even so, we would arrive at costs that, while not disastrous, policy makers should make considerable efforts to avoid.

A Measurement Problem. The studies on which we have based these rough estimates have one particular shortcoming that would lead them to underestimate the economic benefits from scientific research. These studies look at increases in gross domestic product that may have resulted from R&D. If, like the Solow study, they estimate technical change as a residual (the increase in GDP that is not explained by increases in capital and labor inputs), then the higher the estimate of GDP is, the higher their estimate of technical change will be.

Our measures of GDP may be subject to substantial understatement because of their failure to account properly for improvements in the quality of goods and for new products, resulting in an underestimate of the impact of R&D. This failure is particularly noteworthy since a large part of quality improvement and introduction of new goods can be traced back to research and development.

Federal statistical agencies have long wrestled with the problem of measuring quality improvements. Usually the analysis revolves around the price indexes used to adjust output for changes in price, which should take account of changes in the quality of goods. The Bureau of Economic Analysis, the agency that produces GDP data, for example, undertook a special project to determine how much the quality of computers had increased over the years. They figured out how much the price of computing power had decreased[9] and therefore how much the output of computers had been underestimated because of failure to measure price reductions accurately. Computers are an important example of quality improvement but only one of thousands. Statistical agencies do their best to deal with these difficult conceptual issues but lack the resources to address all or even most of them.

How important is this bias? The Advisory Commission to Study the Consumer Price Index gives us some idea of the magnitudes involved.[10] The report gives examples of "new product bias." Air conditioners were widely sold in 1951 but not introduced into the CPI until 1964. Microwave ovens, personal computers, and VCRs were also around for years without being included in the CPI. Today, when there are more than 40 million cellular phones in use, there is still no price index for this high-tech product. The commission estimates that the bias from neglecting new products and quality change is about 0.6 percentage point per year. Of course, the CPI differs in many respects from the index that is used to deflate the GDP, but the issue of quality change afflicts both measures.

Economist Leonard Nakamura of the Philadelphia Federal Reserve Bank has pursued this inquiry as it pertains to bias in the estimates of GDP. He estimates that growth in real output since 1984 has been underestimated by at least two to three percentage points annually, an enormous amount.[11] No one has determined what this change would mean for estimates of the rate of return on R&D, but clearly these estimates would be increased substantially.

R&D as an Investment

Another way to look at the contribution of research is by recognizing that scientific research is an investment: it costs money now and produces benefits in the future. As with any such investment, one can estimate the rate of return to research.

Zvi Griliches pioneered empirical work on the rate of return to investment in agricultural research.[12] He estimated the economic rate of return to agricultural R&D over the 1910–1955 period at about 35 to 40 percent. More recently, Edwin Mansfield has produced a series of studies on the relation between academic research and industrial innovation. One of his key results is that the rate of return to society from investments in academic research is 28 percent. Another of his studies of specific innovations reported social rates of return ranging from 56 to 99 percent.[13]

A 1993 survey article by M. I. Nadiri[14] reviewed sixty-three studies of R&D, most of them pertaining to the United States but also to Japan, Canada, France, and Germany. Based on his review, he concluded that R&D renders, on average, a 20–30 percent annual return on private investment. The return to society overall ranges from 20 to 100 percent, with an average of around 50 percent. These estimates have been cited enthusiastically by proponents of higher science budgets.

Those who have studied the economics of R&D have reached a clear consensus that investment in R&D contributes quite positively to economic growth. Research performed over the years has yielded high rates of return, both private and social. This consensus gives empirical backing for what we will discuss in chapter 5, the role of the government in supporting research. But what does it tell us about the harm that might ensue from cutting science budgets?

Is R&D a Good Investment for Taxpayers? Just as investors want to know how much money their investments are earning, the investors in research (the taxpayer) wonder how much their support of scientific research is "earning" for the nation. An investor with $200,000 in capital and a rate of return of 7 percent, for example, would be getting $14,000 a year. What is the science-knowledge counterpart of that $200,000?

The Commerce Department's Bureau of Economic Analysis

has produced just such an estimate.[15] The BEA has assembled data from the National Science Foundation on the nation's federal and private R&D spending over the years, going back to the 1950s when such data were first available. Using the same methods by which they estimate other capital stocks, BEA applied depreciation rates, assuming that new knowledge, like capital equipment, gradually wears out. BEA's calculations yield estimates of the "stock of knowledge capital." In 1992, the latest year for which these data have been published, the estimated net stock of R&D fixed intangible capital was $1.049 trillion (1987 dollars), of which $842.5 billion was produced by private R&D and $206.8 billion was produced by government R&D.

Now we can make rough calculations of the annual value of the returns to investment in research. Applying a 40 percent rate of return to the $1.049 trillion yields an annual increment to gross domestic product of about $400 billion, or about 8 percent of GDP. (This number is comparable to the $330 figure presented earlier in this chapter that resulted from the "contribution to productivity" approach.) Using a 50 percent rate of return, benefits would be 10 percent of GDP.

Another rough calculation: R&D is 2.6 percent of GDP. Suppose we make a massive 20 percent cut in R&D, which would equal 0.5 percent of GDP. If the rate of return on the lost R&D had been 50 percent, then we would lose 0.25 percentage point of growth each year, a fairly large amount in relation to the nation's long-term growth of about 2.5 percent. Certainly, anyone who could devise a feasible policy that would *add* 0.25 percentage point to economic growth would be an economic miracle worker.

Any stimulus to economic growth has revenue effects. The 1997 federal budget contains a table entitled "Sensitivity of the Budget to Economic Assumptions," according to which a sustained increase in growth of 1 percent a year would increase federal revenues by a total of $420 billion between 1996 and 2002. It would also lower spending by $183 billion. Thus, even an increase in growth of only 0.1 percent per year would lower the net deficit by about $60 billion over seven years.

This analysis, of course, is incomplete. We must also consider what would be done with the money were it not spent on R&D but on some other public investment. Conceivably, the next best alternative might also have a substantial rate of return, which would mean that the *additional* returns from switching from this alternative into R&D might be small. But given the structure of current federal budgets, wherein current consumption and transfer payments far outweigh the investment items, it would seem unrealistic to assume that federal funding of R&D has somehow crowded out some public investment with a comparable rate of return. The other alternative would be leaving the funds with taxpayers. Here, calculating the net rate of return on R&D would mean subtracting out the private rate of return plus the cost of collecting the money from taxpayers.[16]

How should we interpret these results? Earlier we warned about the imprecision of such estimates, and another such warning is due now. What seems reasonable to conclude is simply that we are discussing very large numbers here. Cutting, say, 20 percent of R&D funding will not throw the economy into another Great Depression, but it would eventually create a hindrance to growth comparable to that caused by some other important macroeconomic problems that we worry about—for example, the energy price shock of the 1970s or the costs of overregulation of business. While a cut in the R&D budget would not produce a sudden shock, it would cause a long-term drag on the economy that would eventually reduce the standard of living by a measurable amount.

But as large as these costs are, they may be easily overlooked in everyday life. The lags are sizable; it usually takes many years before even very successful research manifests itself in economic benefits,[17] and likewise it would take some years before these results of research cuts would become apparent.

Moreover, when bad effects did occur, they would not readily be ascribed to a lack of science funding. The nature of the cost is less productivity growth or the *lack* of certain new products, which, being nonexistent, would not be identifiable. In other words, only the scientists (who understand the process) and the economists (who measure it) would have some notion of what had gone wrong. The nation would be missing some benefits of science, but we would not know what they were: a cure for cancer perhaps, a big improvement in energy efficiency, or . . . what?

Because the effects of downsizing are not immediately evident, legislators who wish to reduce R&D funding can do so without causing noticeable harm, except to the "inputs" of research, namely the scientists themselves. This predicament puts scientists in an awkward position, as their attempts to lobby for higher science budgets become an inseparable mixture of social responsibility and self-interest.

The Productivity Slowdown of the 1970s. Another way to consider the effects of R&D spending is to look at how well national R&D performance is correlated with changes in productivity. Can we observe that changes in national productivity occur as a consequence of changes in R&D spending?

Something of a natural experiment took place beginning in the late 1960s and extending into the 1970s, when productivity slowed substantially. Many hypotheses as to the cause of the slowdown were investigated—the energy shock of the 1970s, prior reductions in research spending, diminishing returns to R&D. But the results were inconclusive. Zvi Griliches examined the relation between the productivity slowdown and changes in R&D in his 1994 presidential address to the American Economic Association. He ascribed his inability to find a definite link between productivity and research to a lack of good relevant data.[18]

Would Downsizing Be Costly?

At face value, the empirical work makes a strong case for federal funding of science. Few, if any, federal programs are backed by such a mountain of favorable economic evidence.[19] This point deserves emphasis. The great majority of federal spending goes not for investment but for transfer payments—virtually a zero-sum game insofar as such spending affects economic growth. Of the programs that produce or purchase goods and services for public use, few can compare favorably with R&D as a good investment.

How Much to Spend on Science? The empirical evidence gives little guidance, though, as to whether the science budget should be $50 billion or $150 billion. Even the information that the *average* rate of return on science is 50 percent is not proof there should be more; maybe the *marginal* rate of return—that which applies to an increment of R&D funding—is much lower than 50 percent. This average-marginal problem applies equally to assessing budget cuts. That is, if there is a 10 percent cut in the budget for R&D, will the benefits also decline by 10 percent, or by more or less than 10 percent? The following considerations suggest that the estimated magnitude of benefits from science funding must be qualified in several important ways.

Eliminating Low-Return Projects. Could enlightened policy target the lowest-rate-of-return projects for elimination? After all, funding agencies should have a fairly good idea of how to rank the projects they fund. Anyone who has read numerous research proposals knows that many, even among those that are funded, will not yield a 50 percent rate of return or anything close. So can we drop the least promising 10 or 20 percent of R&D and hardly miss it?

The problem is that it is extremely difficult to predict the results of a specific research project, let alone rank it among others according to the ultimate worth to society. Research is just too uncertain and elusive. Moreover, the budget allocation process, as it is now constituted, does not even attempt to do this except very indirectly. Many research proposals undergo a peer review, which looks not so much at social benefits but at whether the project would be "good science." There is some correlation here, of course, as "good science" is a requisite for social benefits while "bad science" is without benefits. Nevertheless, it would be unrealistic to assume that tighter budgets would automatically target the least productive research. At the aggregate level, the existing process has no effective way, other than by political judgment, of deciding whether $50 million should be taken out of the space shuttle as opposed to AIDS research or something else. Political considerations for funding, such as traditional pork barrel procedures or pressure-group lobbies, have little to do with social benefits. Thus, while the marginal losses from rationally cutting science spending might be minimal, the word *rationally* is the sticking point.

Federal versus Corporate versus University Research. The discussion in the previous paragraph applies mainly to federally funded R&D. What about relative returns among the categories of spending, such as public versus private? Most of the economic studies cited deal with aggregate R&D spending, or to corporate R&D, whereas the downsizing with which we are concerned is in federal funding.

Although federally funded research is often said to be less

productive than private sector research, variation within each category is wide. Some federal research is excellent; some is ineffectual. Lichtenberg, for example, found that government R&D spending had a much lower net impact on productivity than did privately funded R&D.[20] Linda Cohen and Roger Noll found many expensive failures among the federal projects they studied, but most of those were part of one particularly sorry episode in federal science history, namely the misguided attempts inspired by the "energy crisis" of the 1970s. Most of these projects were performed at government-operated facilities, not at universities or in the private sector.[21]

On the brighter side, Edwin Mansfield's estimates of very high rates of return apply specifically to federally funded research performed at universities. A recent study of patent citations by Francis Narin of CHI Research found that more than 70 percent of the scientific papers cited on U.S. industrial patents came from federally funded science, the rest coming from industrial research.[22] The study found that 52 percent of the sources of the cited scientific papers were academic, and just 11 percent were from government labs. Thus, even if research at universities is largely "fundamental" in nature, some of it eventually contributes very heavily to practical applications in industry.

Federally funded research varies greatly from program to program in its contribution to society. Unquestionably, some federal research projects have a low or even negative rate of return. To the extent that such projects are common, correctly targeted cuts in federal funding would be less harmful than indicated by many of the rate-of-return studies. Some argue, for example, that some of the federal laboratories are too big and could be downsized without great loss, owing to changes in the importance of their weapons development mission and to inefficiencies in their operation.[23]

Long Lags Before Benefits Occur. Because it often takes many years before the benefits of research are realized, perhaps, the argument goes, we could reduce science spending and not feel any pain, since the costs would be passed on to future generations. This view calls to mind the image of a long conveyor belt that keeps delivering things at the end well after those items have ceased being placed on the conveyor belt at its beginning. This image is misleading, however, because the fundamental research of years ago does not simply dispense its benefits automatically; today's researchers must draw them out.

Consider the research on prime numbers that began centuries ago. Around 1859, Riemann developed the "zeta function," which encodes information about how primes are distributed among other integers. Only recently has it been discovered that the zeta function also provides a way to simulate the behavior of complex atomic systems described by quantum physics.[24] What had been pure mathematics is now being applied to the physical world. Thus, the work of Riemann and his predecessors is still bearing fruit, but only because the tree planted so long ago is still being tended. Paul Romer describes this quality of research in economic terms: new knowledge is a form of capital to which diminishing returns do not apply, as it can be used over and over without being used up.[25] The "using" part of the process is immediately affected by funding reductions.

Thus, if we cut today's budget, we reduce the nation's current ability to make use of yesterday's discoveries. Even to the extent that we could cut research now and not feel the pain for ten years or so, does anyone want to make intergenerational burden shifting the cornerstone of our science policy?

The "Accelerator" Effect and the Human Dimension. In the previous chapter, I brought up the accelerator effect, whereby a change in the funding for science is likely to cause a much larger change in the demand for scientists. This effect can also incur abstract costs, such as a drop in morale and the loss of potentially good young scientists to other professions. The resulting aging of the scientific work force, and the possible loss of creativity, is discussed in chapter 8. The effect of these intangible costs, all else constant, is that the science enterprise is probably hurt out of proportion to any loss in funding.

Replacing Cuts in Federal Funding. Would private funding replace lost federal funding? This is one of the key questions of downsizing. It is one of the criteria by which we should judge whether a program should be funded. Obviously, very little if any social benefit is to be gotten from funding research that would have been done anyway. Less obvious is the question of whether federal policy has adequately provided incentives and eliminated disincentives to the private sector's funding of research.

Has federal funding, over the years, displaced a lot of private funding that would have take place in its absence? If so, then we might be able to reduce federal funding without pain, since private funds would pour in to take up the slack. It seems probable, however, that private funding would not come close to filling the void.

Private foundations are sometimes mentioned as alternative funding sources. They contribute to science, but not on the scale that the federal government does. Foundations currently direct only 4 percent of their grants to scientific research, just $250 million in 1993.[26] Even doubling this figure would cover only a fraction of the prospective federal shortfall, an indefinite figure but surely well into the billions of dollars per year. Given the many other demands on their resources, private foundations are unlikely to move into science funding on the scale required to close the gap.

Private companies are in a better position than foundations to support more of the nation's research, but would they? Industry funds research that is expected to result in profits sooner or later. While industry does fund some research classified as basic, it is highly unlikely that industry would have funded much of the fundamental research that the National Science Foundation supports, or that of the National Institutes of Health, or the space program, or whole fields of science, like anthropology or astronomy, that do not produce profits. Some military research would get company funding if it helped to land weapons contracts, but the vast majority of the Pentagon's budget is mission-oriented research that would be unprofitable to industry.

Any evidence? Christopher Hill of George Mason University has found a positive correlation between federal and industrial spending on research, implying that federal funding cuts might induce industry cuts.[27] My reading of his data is that companies are unlikely to fill the gap left by the government, but neither is there persuasive evidence that federal cuts would cause private research cuts. In my view, private and public funding are mostly independent of each other, although they can be affected by the same factors (the health of the economy, for example), so while they are statistically correlated, neither actually influences the other.

Another possibility for replacing federally funded research would be the research carried on in other countries. While any particular foreign response to reduced U.S. funding is unlikely, the fact is that R&D is growing worldwide. If the United States is able to take advantage of this increased stock of knowledge, it would make up for a part—and likely just a small part, given the nature of international transmission of scientific knowledge—of what is lost through U.S. downsizing.

Finally, would universities be able to replace lost federal funds with their own? Not to any useful extent, chapter 8 argues. Federal funds supported 60 percent of university research in 1995, and alternative funding sources are either nonexistent or unlikely to be diverted into research.

Conclusion

In sum, there are several qualifications to the determination that research has a high rate of return. It is hard to say whether the output of science would decline more or less than in proportion to funding cuts. Factors work in each direction. Perhaps the most important consideration to be drawn from this discussion is that it should be possible to target budget cuts to the least productive forms of research, thereby ameliorating the costs of downsizing. If we cut out the worst programs, we will not lose much. This strategy, however, presupposes a system of setting priorities for federal funding that, unfortunately, does not exist.

This chapter has reviewed evidence that the economic benefits of R&D have been large. Even the most favorable interpretation of the evidence, however, does not prove that science deserves a blank check. There are too many uncertainties. One problem is that in searching for measurable effects of science we have taken an excessively narrow view: we have looked at what was most easily measured (changes in GDP, for example). The next chapter looks at the impact of scientific research in other, broader contexts and considers which of those impacts might be affected by downsizing science budgets.

4
A Broader View of the Benefits from Science

ECONOMIC BENEFITS FROM science are quite substantial, as we saw from the evidence in chapter 3. The question, then, is not whether decreasing scientific research will reduce those benefits but rather by how much. The results of the economic studies are subject to any number of caveats, and economists have some difficulty in using those results to predict precisely what would happen in the face of federal cutbacks. One can fully accept the economists' findings and still believe that science budgets could be trimmed around the edges without much loss. In addition, some skeptics of the accuracy of the estimates would be comfortable with much larger downsizing. Thus the argument stands at this point.

Let us now turn to the broader results of research, since the strictly economic effects as normally measured may yield too narrow an inventory of benefits. We need to ask the more general question, How does science contribute to the quality of life? Yet even this approach may be too narrow if it looks just at individual pieces of research. Scientific research is far more extensive than a catalog of projects and their cumulative results. It is at once a complex enterprise and an organic whole. *Synergy* may be a buzzword, but it aptly describes the way the elements of science react with one another and with the quality of life.

More than Economic Benefits

First, let us consider just how limited the measures of economic benefits really are. The fundamental product of research is knowledge. Some of this knowledge may have absolutely no practical use, other than being truth about the world that someone wanted to know. (Of course, many "impractical" pieces of knowledge have unexpectedly turned out to have some important application.) Such "useless" knowledge, however, may well be valued by society, just as Vermeer paintings and Brahms concertos are valued.

At the other end of the spectrum of practicality is new knowledge used to produce goods at lower costs or to produce brand-new goods and services. To measure these effects, the Solow-type studies look at increases in output that are unaccounted for by increases in inputs like capital and labor. Output is measured by gross domestic product, if the study covers the whole economy, or by some measure of output of the industry or firm if the study is more limited. Therefore, if the knowledge resulting from research does not affect GDP, then it is not captured in this type of measure.

Although these economic measures can be stretched conceptually to cover many kinds of benefits, in practice data limitations dictate that only some of the benefits of research are in fact measured. Several of the studies cited in chapter 3 attempt to measure the benefits of research spending by calculating changes in the profitability of companies that performed the research. Others estimate benefits through "consumer surplus," a measure of the monetary value of having available a new good or an already existing good at a lower price. If a result of research does not affect profits, or the price and quantity of a good about which we have data, then it is beyond the scope of such studies.

In the previous chapter, we discussed how a data problem—the underaccounting for new products and quality change—biased downward the estimates of economic benefits from research. Alternative ways of looking at the benefits of science may be called for: instead of simply totaling up economic benefits, let us look at the vast changes in our way of living that have come about gradually when viewed year by year but massively over time, through technical progress.

Between the two extremes of practicality, we have a broad

35

class of new knowledge: that which confers benefits that people would be willing to pay for but that escape measurement in our economic data. Let us consider some examples.

Improved Health. Improvements in health and longevity are among the largest and most obvious benefits of scientific research. Since 1890, life expectancy in the United States has increased by about thirty years, and infant mortality rates have fallen by 90 percent. Indeed, research aimed at improving health is doubtless the most popular form of publicly supported research; it consumes nearly a third of the federal nondefense research budget. In federal budgets of the past twenty years, funding for health-related research grew rapidly and, despite setbacks here and there, is likely to be the most impervious to downsizing of any field of science. The Department of Health and Human Services currently spends around $12 billion annually on health-related research.

Not that all health improvements can be attributed directly to science, though. Improvements in nutrition resulting from economic growth, for example, have had a big role in lengthening life spans and reducing the incidence of some diseases.[1] Richer people are healthier. Nevertheless, the benefits that come directly from research are enormous: new drugs, new therapies, new diagnostics, and so forth. Better health over the years has not been just a matter of taking new pills. By discovering the basic causes of disease, scientific research has given people the information they needed to take steps to fend off diseases with better sanitation, hygiene, nutrition, and other now mundane precautions.[2]

Some benefits of improved health are reflected in measures of improved productivity. Healthy workers are more productive than sickly ones. But more than half of the population is not in the labor force (mainly those who are too young or too old), and the improved health of those 140 million people needs to be included in the benefits, too. Moreover, the benefit of someone's simply feeling better because he has avoided illness is not part of GDP, except insofar as the benefit was purchased as part of health care.[3]

Recognizing Hazards. Science allows us to understand better the world's hazards to life and limb. Cigarette smoking, for example, was long known to be unhealthful, but only in general terms. After years of scientific research, the risk has been quantified; we know fairly well which diseases smokers are likely to get, with what probability, as a result of how much smoking. With this information, more people have taken the dangers of tobacco seriously, and smoking has dropped considerably, as has the incidence of the associated diseases.

This same process has been repeated for innumerable health and safety threats, with benefits too large and diffuse to calculate. Better understanding of the true costs of environmental degradation has led to beneficial antipollution measures.[4] To take one current example, we need to understand (1) whether "global warming" is occurring; (2) if so, what is causing it; and (3) whether any of the expensive regulations proposed to deal with it are likely to work and be worth the cost. Good science is at the core of each of these questions. Moreover, without the decades of previous work, in various fields, by scientists who were not even thinking about global warming, we would not now have the capability to study this issue intelligently.

The process, of course, does not work perfectly. Scientific evidence is sometimes drowned out by the shriek of junk science, scare journalism, and always-ready-to-sue-on-contingency lawyers. Often these scares are overblown, or just plain wrong. Sometimes regulatory authorities appear either to underrate or to overrate various health or environmental risks. A current example is EPA's unpersuasive scientific justifications for its proposed expansion of Clean Air Act regulations.[5] Thus, the nation often fails to get the full benefit of scientific methods as applied to potential risks. Nevertheless, the point is that science has made available to us knowledge that will—if we use it right—save lives and improve living standards.

Saving the World's Natural Environment. The biggest danger to the world's natural environment may be low-yield agriculture. The world's population will probably exceed 8 billion within thirty years, and the great majority of these people will be rich enough to demand large quantities of resource-costly foods: meat, milk, eggs, fruits, and vegetables. Thus, it is estimated that the world's farm output must increase by at least 150 percent and may need to triple. The "green revolution" made possible by agricultural research has brought higher yields in recent decades, but its technologies are already fully exploited by much of world agriculture. In many countries, agricultural output is being increased by putting more land under cultivation, with destructive consequences for wild habitats and wild species. The only food strategy likely to protect the world's remaining forests and wildlife is to increase sustainable crop and livestock yields even further, and that increase will require more biological, environmental, and agricultural research.[6] Thus, estimates of the rate of return from agricultural research that consider as benefits only the market value of higher crop yields are too narrow.

Benefits from the Social Sciences. The social sciences get federal money—several hundred million dollars a year. What sort of benefit comes out of that research? Economics, sociology, political science, and anthropology are the principal social sciences. Although the social sciences use quantitative methods similar to those of the natural sciences, they are not always accepted as full members of the science club, and they are sometimes called the "soft" sciences in contrast to the "hard" natural sciences.

Benefits from social science research are likely to be even more abstract, elusive, and hard to measure than those in the natural sciences. During the congressional debate over the legislation that established the National Science Foundation in 1950, there was disagreement over whether to include any money at all for the social sciences. While some argued that social science research would provide useful support for the army's psychological warfare program, among other things, the social sciences were excluded mainly for reasons of economy.[7] Not until 1954 did the NSF establish a program in the social sciences. In 1993, the federal government spent $338 million on social science research at academic institutions.

Since spending on social science research is contained in the aggregate of national R&D, we might simply assume that it produces benefits in the same ratio as other fields of research.[8] But what exactly are these benefits, and are they measurable? Consider the application of economic research to public policy; the uses of research on monetary theory and policy are an example. Thomas C. Melzer, president of the St. Louis Federal Reserve Bank, said:

> It is my experience that Federal Open Market Committee members make decisions mainly based on their core beliefs about the effects of monetary policy . . . . These core beliefs are obtained essentially by talking with econo-

mists—either directly or through books, media, friends, and colleagues. And economists' core beliefs are formed, ideally, through the best available formal theoretical and empirical research on these questions.[9]

The Federal Reserve Board's Open Market Committee determines the nation's monetary policy. Its decisions can and do change national income and wealth by tens or even hundreds of billions of dollars per year, along with millions of jobs. This influence was demonstrated spectacularly by the consequences of the Fed's bad decisions that helped bring on the Great Depression and the resulting years of lost incomes and human misery.[10] If one accepts the premise that research since that time has led to better monetary policy that has prevented or ameliorated some potential recessions, one could conceivably estimate, or at least bracket, the payoff to economic research on monetary policy, probably arriving at very high rates of return. Good monetary policy also keeps inflation in check, which is essential to economic and social well-being. The Federal Reserve spends considerable sums for research to guide its policies and operations.

Other fields of economics have yielded public policy benefits (to the extent that their recommendations have been followed): free trade is beneficial, governmental regulations are often costly, and free markets work better than state planning, as well as many corollaries of these principles.[11]

It is not clear whether all fields of social science, particularly those less directly applicable to public policy, can reasonably claim such large payoffs from their own research.[12] It is always possible to take credit for enlightened social policy, but it is much harder to establish a clear connection between successful policies and prior research. Indeed, one would have to net out the harm from bad policies resulting from following recommendations based on faulty research. One must also note that much of the most influential social science research was performed without any federal support.

A Multidisciplinary Example. While dividing science into its component fields is useful for some purposes, science has a way of obtaining knowledge common to all its fields. Ideas and knowledge flow from one field to another, often ricocheting around in odd ways, and leading to still newer knowledge. Often it is impossible to unravel the multiple threads of work that brought about one new invention or improvement in knowledge, and it is often equally futile to try to determine the relative roles of federal and private funding, or of "fundamental" versus "applied" research.

To show just how broadly our concept of benefits must be to appreciate the contributions of science to health, consider the case of "El Nino/southern oscillation" (ENSO).[13] The interaction of medical research, meteorology, oceanography, satellite communications, mathematics, and archeology is evident in this example.

The term *El Nino* comes from the observation by Peruvian fishermen that in some years the waters off the Pacific coast of South America are unusually warm at Christmastime, with consequent reductions in the numbers of fish. Independently, a large-scale oscillation of atmospheric pressure between the eastern and western tropical Pacific Ocean was discovered, which was linked to El Nino.

Eventually the two events were found to be aspects of a single process, whereby moist, warm air rises from the surface of the Pacific to disrupt the jet stream in the upper atmosphere, where weather conditions develop. This interaction has all sorts of global repercussions. In addition to effects on marine ecology, consequences include drought in northern Australia, Indonesia, the Philippines, Brazil, and southern Africa. Effects extend to the United States: mild winters in the Northeast, heavy rains in California, and a lower chance of hurricanes along the Atlantic coast. Health researchers have linked ENSO to outbreaks of cholera and other infectious diseases in Latin America and Bangladesh, which only in recent years were found to be linked to certain types of plankton blooms that grow in the warmed waters. Thanks to satellite pictures and other monitoring of the oceans, conditions presaging a cholera outbreak can be detected. Mathematical models have been developed that have made accurate predictions of El Nino conditions. These models may be useful in reconstructing prehistoric changes of climate, using research results on chemical tracers, temperature bands on ancient corals, and other newly discovered archeological clues.

All of this suggests potentially high economic payoffs from research resulting in better weather predictions that would permit people to prepare for bad weather and avoid part of its damage. In crop savings alone, better predictions of ENSO effects could be worth $240 million to $323 million per year, according to the Agriculture Department.[14]

Other Unique Aspects

Economists continue to refine and reinterpret their conception of how technical change contributes to economic growth. While the Solow approach showed a way to account for the effects of innovation, later work has tried to present a more integrated model of both cause and effect. One strand of research seeks to determine the extent to which convergence in output per worker among nations is happening and what is causing it. It is increasingly accepted that some nations (or regions) grow faster than others not because of their endowments of capital or because they attract capital from abroad but because the characters of their economies and their societies are hospitable toward innovation. It may be something specific, like favorable tax rules, or something ingrained in society, like the element of "trust" that Francis Fukuyama discusses in *Trust: The Social Virtues and the Creation of Prosperity*.[15] The general term for all this is *endogenous growth*, of which technical change is a cause; it refers to economic growth that happens not of its own accord but because "someone has done something."

A healthy scientific enterprise contributes to the process of growth by its individual projects as well as by improving the atmosphere for innovation. Richard Nelson and Nathan Rosenberg use the term *national innovation system* to refer to the "set of institutions whose interactions determine the innovative performance . . . of national firms. . . . The 'systems' concept is that of a set of institutional actors that, together, plays the major role in influencing innovative performance."[16]

Another way that research contributes to the whole of society is through its role in education, or the accumulation of "human capital." One of the largest benefits of public spending on scientific research is the resulting improvement in skills and knowledge of the work force. People educated as scientists and engineers are made more productive, able to apply the scientific method to solve problems that are nothing like the subjects they studied during their formal training. Research is itself a form of education. It is increasingly recognized that graduate students employed as research assistants are receiving a unique complement to their traditional classroom training. The 20,000 or so postdoctoral fellows, who continue on in a more senior apprenticeship, are also expanding their personal skills and store of knowledge.

Questionable Benefits

Sometimes proponents of science funding tend to exaggerate by counting benefits but not the offsetting costs. Local proponents of some new research facility, for example, often claim that some number of jobs will be created. More accurately, jobs would be shifted from whatever locality loses out in its bid for the project. If it is just a question of building the facility in a certain place or not at all, then we need to consider the benefits of leaving the money with taxpayers, as well as the costs of taxing it away from them.

The benefits discussed in chapter 3 were not the sort of macroeconomic benefits or "multiplier effects" that are sometimes claimed for government spending. A 1996 story in the *Boston Globe* proclaimed that "Massachusetts could lose tens of thousands of jobs and billions of dollars in economic activity if federal spending is reduced."[17] This might be true for Massachusetts, but from the national point of view, jobs are not really "created" by taxing or borrowing to build a laboratory (or post office or football stadium). They are shifted from wherever in the private sector the funds might have been spent had the government not absorbed them. Local politicians can help their constituencies by attracting federal research dollars, but the net macroeconomic benefits to the nation of local spending are nil. It is the research itself, not the spending on it, that benefits the nation.

Conclusion

These examples all portray substantial benefits that are not fully captured by the economic measures discussed earlier. Quite clearly, more such examples could be given, all to demonstrate that while the economic benefits of science are quite large, they are far from being the whole story. Therefore, we cannot rely solely on the economic analyses, particularly if we interpret them as showing that downsizing science would be fairly painless. Many benefits beyond the economic ones stand to be attenuated.

The discussion in this chapter tells much about why the nation supports science. We fund health research lavishly without bothering to measure the economic benefits that will someday accrue. The source of support is a general perception that health research will be good for us. Quite simply, we like it. This is also true of other fields of science—we like anthropology not for its payoff, which at best is obscure, but because it tells us something we want to know for its own sake. Astronomy and space exploration are more examples of research we want to do, never mind balancing costs and benefits too carefully.[18]

Certainly, popularity is an important factor in science funding. In a way, public approval is a measure of the output of research. Therefore, we should look to public perceptions of science to understand just how strong public support is likely to remain, our subject for the next chapter.

5
A Vital but Limited Role for Government

SCIENTIFIC RESEARCH PRODUCES many benefits, but so do many other activities. Why should government be involved in science, and what are the limits to its role? If federal budgets are to be tight, can we simply cut out the least justified programs without losing much social value? To answer this question, we need to develop criteria for governmental intervention in R&D. These criteria follow directly from the economic concept of "public goods."[1]

A pure public good is one that is "nonrival" and "nonexcludable." That is, one person's consumption of the good does not diminish the ability of others to consume it, and the producer of the good cannot prevent others from using it. National defense is a perfect example, and consequently there is no question that the government should be responsible for defense. In fundamental science, the "goods" are primarily information, like the equations of Newtonian mechanics and the basic understanding of DNA. Such goods cannot be used up, and no one can be kept from using them. The benefits spill over to every continent and last for centuries.

Without an operational market for broad-gauged research results like these, no one has a sufficient incentive to produce them on a large scale. The discoverer of new information has no way of collecting payments from users. In the absence of federal funding, "too little" research would be performed. Some amount of

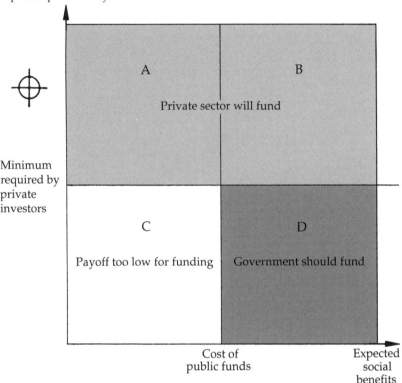

FIGURE 5–1
PRIVATE AND SOCIAL BENEFITS FROM RESEARCH AND DEVELOPMENT

SOURCE: Author.

public-good–type research would be done, just as it was in times before large-scale governmental involvement, but only enough to satisfy private incentives (like personal curiosity or altruism). Some commercially profitable research also delivers social benefits. But we are still left with the category of research that would provide substantial social benefits but whose private benefits are insufficient to attract support. Therefore, without governmental support of these projects, the nation would underinvest in fundamental research.

In broad terms, figure 5–1 illustrates the conditions under which federal funding is justified.[2] The private sector supports R&D if its prospective returns exceed some minimum, which occurs in quadrants A and B. In quadrant C, neither private nor social returns justify financing the projects. Government funding would be justified only for projects in quadrant D, which the private sector would not find sufficiently attractive. Projects in quadrant B have high social returns, but why should the government spend money on them if they will attract private funding? This exposition is oversimplified, particularly the idea that benefits from R&D can be foreseen or measured with much precision, but it lays out the issues that policy makers must consider.

Reasonable Limits

Sponsors of federal science programs often apply an overly simplified public-goods criterion to justify what they propose to fund. But there are two important caveats.

First, excludability can sometimes be established by means of property rights. People can be excluded from using the good in question or can be made to pay for its use. Before spending federal money in a particular instance, one needs to examine whether the "nonexcludability" circumstances can somehow be changed. The president's 1995 economic report expresses it this way:

> Economists have estimated that . . . innovators typically capture less than half the total social returns to their investments in research and development. In short, the difficulty of establishing and enforcing property rights to new ideas reduces the economic incentive for private companies to invest in a socially and economically optimal level of R&D. Bolstering that incentive is therefore an important efficiency-enhancing function of government.[3]

In other words, establishing incentives and setting up an efficient regime of property rights may induce private individuals to undertake beneficial research and relieve the taxpayer of the burden. In the classic "tragedy of the commons," where a community pasture is ruined through overuse, the market solution is to charge grazing fees. In cases where the common resource is not easily owned (fish in the ocean, for example), the government may try to approximate the property-rights ideal by auctioning licenses that allow the licensee to catch a certain number of fish. If property rights can be established and excludability enforced, the government has no basis for involvement, even for nonrival goods. In figure 5–1, some projects can move from quadrant D to quadrant B.

By extending the scope of patent protection and thus enabling inventors to reap greater financial rewards, the government can encourage research in certain areas. There is, though, a problem of balance. If patent law is too weak, then the incentive to look for patentable discoveries will be weak. If patents are too strong and provide too much protection, patent holders would charge too high a price for the use of knowledge that is essentially costless to use, and so the social benefits would be too small. This is a classic dilemma of science policy. Another example of an inefficient institutional constraint is antitrust laws that prevent firms from forming R&D joint ventures. This problem has been largely solved through remedial legislation.

Economists, not just politicians, are sometimes too quick to declare a good to be "public," when in fact a market-induced action will provide the good. A leading economic text used to point to the lighthouse as the archetypal public good. A lighthouse benefits all ships, but no one will build one because others cannot be excluded from using its light and hence cannot be made to share the cost. Government, the text concluded, must build the socially beneficial number of lighthouses. But reality is more complex. Nobel Prize economist Ronald Coase showed that in England lighthouses were in fact provided privately for many years without government intervention, by means of an association of shippers.[4] Creative entrepreneurs have produced other supposedly public goods.

The second caveat: sometimes the government just cannot do the job. Government may face technical problems that it cannot solve, or it may fail through inept management. The Energy Department's management of its national laboratories, for example, is widely criticized as inefficient.[5] In such cases, the government may have simply chosen the wrong institutional structure (federal management versus contracting out or privatization) for an inherently worthy purpose. Or maybe the problems of government inefficiency are insurmountable.

All costs must be counted. Proponents of federal funding (not just for science) rarely consider the costs of collecting taxes, which include not only the cost of the Internal Revenue Service but also the excess burden on the economy, in the form of perverse incentives. If higher taxes dissuade people from working or investing, for instance, that cost needs to be included, as does the administrative cost of running the IRS. Studies indicate that this excess burden is around twenty cents on the dollar, which means that a federal program needs to produce more than a 20 percent rate of return to justify its existence.[6] In figure 5–1, the vertical line separating quadrant A from quadrant B (which delineates the minimum qualification for federal funding) may be far to the right.

Sometimes, then, it is better to ignore a "market failure" than to replace it with a government failure. The record of government intervention in technology and elsewhere suggests that the burden of proof should lie with those who propose to extend government's role.

Criteria for Evaluating Programs

Based on the analysis so far, the necessary conditions for legitimate government intervention are these:

- The research must have a reasonable chance of producing benefits that outweigh its costs. These must be broad social benefits (rather than benefits just to a particular firm).
- The research would not be performed in the absence of government support.
- The federal agency must be able to fund or perform the research efficiently.

These criteria apply equally to research grants, contracted research, and "mission" research performed by government laboratories. Let's take a closer look at each of these criteria.

How Great Are the Social Benefits? In some ways, the reputation of scientific research is too good. Federal agencies tend to act as if more research is always better and resort to subsidies of one sort or another. Such a subsidy may be meant as a way to "improve the competitiveness" of some group of companies, but if the benefits remain primarily with the companies that perform the research, then the public-goods justification does not apply.[7] The benefits of such programs become even more disconnected when multinational firms are involved. Is General Motors (with extensive overseas operations) a domestic company, and is Toyota (with big factories in the United States) a foreign company? Again, identifying the true social rate of return is the key consideration.

Subsidizing science education poses a problem in defining social benefits. Shouldn't Ph.D. students pay their own way, one might ask, since they are acquiring skills and knowledge that they can sell during their careers? Indeed, many of them do finance their own training. The justification for federal fellowships may be that they are part of the total federal science package and that the government ultimately gets more from its R&D dollar by ensuring an adequate supply of scientists rather than depending on a labor market that may not be fully efficient.[8]

Displacement of Private Spending? Displacement is a consideration with many federal programs besides R&D funding. The social insurance functions served by fraternal and craft organizations, for example, were almost completely displaced by social security and other income security programs.[9] Federal loan programs intrude into private sector credit markets. Medicare to some extent pays for what people would have paid for themselves. Thus, a federal program may look good in isolation, but if it simply replaces what would have occurred anyway, then its rationale disappears, unless the point was simply to transfer income from one group to another.

By the criterion of putting broad social returns first, the most obvious candidate for funding is a research project where the social benefits are expected to be large and the private benefits are small. The most obvious noncandidate is the opposite, a project with no social benefits but with positive private benefits. In practice, research usually yields some of each kind of benefit. It is tempting to subsidize anything that has substantial social benefits, even if the private benefits by themselves are large enough to attract funding. To restrict federal subsidy to projects that would not be done without federal support is the consequence of this line of argument, but then the government would, in the end, subsidize commercially nonviable projects, a hard concept for taxpayers to accept. If the social benefits were initially overestimated and turn out to be illusory, then taxpayers are left with a turkey.

Suppose a potential project is likely to have both high private returns and high social returns, and a corporation decides to proceed with this research. One might argue that a subsidy would induce the firm to undertake the project faster, or on a larger scale, and thereby produce even larger social benefits. But with this approach, the opportunities for subsidy would be immense. One could justify subsidizing almost everything. Moreover, getting the subsidy "right" would take more fine-tuning ability than any government agency has consistently displayed. Nevertheless, as we shall see later, this approach appears to be the one taken in some current federal programs.

For "mission oriented" research, displacement is usually not an issue. Much of federal research is directed to the specific needs of a federal agency, such as defense or public health, and would never have attracted private funding. Just as mission research does not displace much if any private research, the private sector would not pick up much of the mission-research slack left by a downsized federal budget.

Displacement is more relevant to programs that support commercial technology. The Commerce Department's Advanced Technology Program, for example, gives matching funds to groups of companies for their own research. Since to get the awards the companies have to show that the research results will be commercially useful, it is plausible that many of the proposed projects would be attractive enough to receive private funding on their own. But if the government announces that it will fund the research, then companies will naturally accept the money offered.[10]

Efficiency of Research. How efficiently can research results be delivered? It is one thing to justify an action in the abstract and another to set up a system that delivers the goods efficiently. The *Washington Post* is a great newspaper, but if the delivery boy throws it onto your roof half the time, then its benefits to you are much diminished. Similarly, however beneficial the public good may be in the abstract, it must be delivered before we can count the benefits. Inefficiency in the delivery process reduces the effective rates of return, which in chapter 3 were crucial in estimating the harm from downsizing science.

Just how efficient is the government in delivering benefits from the R&D it funds?[11] The issue here goes well beyond the simple concept of "waste, fraud, and abuse" by which inspectors general evaluate agency performance. The first hurdle comes in designing a program, which entails determining what sort of project is to be supported. Project designers have great latitude here, given the virtually infinite number of possible approaches to research. In general, is it better to support fundamental science at universities or applied technology through the Commerce Department's Advanced Technology Program? (All else equal, fundamental research, as opposed to technical development, is much more in line with the precept that spillover benefits are paramount.)

What field of science should be supported—more biomedical, more astronomy, or more basic physics? Here the federal government has erected labyrinthine processes for setting budget priorities across the complex federal structure. Expert views differ on how effective this structure has been in recent administrations; some say it is too complex and needs more focus on the big issues.[12] Clearly, this mechanism will face new challenges as the budget declines instead of growing.

Which of the projects competing within a field should be supported? At this time, much of that decision making is done with the help of outside experts (peer review) drawn from the scientific community. Does the funding agency pick the projects most likely to result in the highest social rates of return? How confident are we that social benefits are maximized when it is scientists themselves who advise on the disposition of science funds to the fields they represent? The time-honored peer review system is generally believed to work well in most respects, and no one has devised any vastly different system that would be more efficient. Nevertheless, when a body of scientists is given broad authority to allocate funds within their own field, questions arise about whether their objectives match society's. Does "the best science" necessarily equate to the largest social benefits?

What about the cost of the peer review system? Peer reviewers must be first-rate scientists whose time is quite valuable. The process of writing proposals is also time consuming, and as success rates decline (because of lower science budgets) the cost of the whole process, per dollar of grants awarded, will increase. Would some form of institutional support contribute to greater efficiency of making awards? Or is there some way to reward actual research success rather than skillfully written proposals?

What about the federal labs, where the government is the funder and performer of the research? Here we need to consider bureaucratic inefficiencies, to which numerous review panels have attested, but also the missions themselves. If the wrong research is performed efficiently, the rate of return is still low.

In sum, in a consideration of funding reductions, it makes sense to target programs where private funding might very well take over. When possible, matching grants should replace full support grants, stretching the funding and encouraging private support.

Contrary Views

Some question any role at all for the government in promoting science. Economist William Niskanen, for instance, asks how the United States survived and prospered for so many years before all this government help arrived.[13] His skepticism is based partly on his interpretation of the studies reviewed in chapter 3, where the evidence fails to persuade him fully that science has a big payoff. Furthermore, he sees federal support as largely displacing private research. His preferred public policy is to augment private R&D expenditures through the tax system, by a tax credit for research and experimentation, one larger and more inclusive than the one that has been used, off and on, since the early 1980s. He also suggests matching grants as a means of inducing more government research without putting government completely in charge of selecting specific research projects.

Terence Kealy launched a longer, more detailed attack on governmental support of science in *The Economic Laws of Scientific Research*.[14] In my view, this work is flawed by illogic, and I mention it here only because of the attention it has received in the debate over science budgets. Kealy's argument is that science would be adequately supported by private means without governmental intervention. One problem, though, is that nearly all his examples of private funding go back many decades, to times when science was cheap (no colliders or genome mapping required) but were nevertheless carried out on a tiny scale compared with today. His modern examples are carefully chosen cases of failed federal projects, and his older examples are all great successes. He does not discuss the earlier era's research opportunities lost for lack of sufficient support, nor does he give adequate credit to the innumerable good results from federally supported research. In any case, modern science is done on a scale vastly larger than at any time before 1940, and it seems clear that regressing to that earlier cheap-and-sparse science would deprive the world of enormous potential rewards.

Another challenge to the funding orthodoxy is the radical approach of Charles Murray's *What It Means to Be a Libertarian*.[15] Murray presents the same definition of a public good as this chapter does, but he limits his ideal government to only the most necessary interventions. He calls the acceptance of the public-goods argument a "slippery slope," on which one can easily slide into facile justifications of a huge welfare state. He would wipe out all subsidies for scientific research, leaving only enforcement of patent law as the sum total of science support.

I would agree that, governmentwide, the public-goods rationale is applied far too easily and is often little more than camouflage for political payoffs to favored groups. Nonetheless, I believe that fundamental scientific research is among the most legitimate of public goods, so long as the criteria developed in this chapter are met.

Conclusion

This chapter has presented the economic case for federal support of science. In general terms, the case is strong. But it is not a blanket justification for everything the government might want to do by way of subsidizing research. The public-goods defense of federally funded research carries with it certain principles that determine what is and what is not a legitimate use of federal funds. To the extent that a particular program diverges from these principles, its demise would not be particularly costly to the nation. Its role should be limited to research that (1) promises to deliver broad social benefits, (2) can be delivered efficiently by government, and (3) would not be performed in the absence of support. Subsidizing firms to perform research that they would have done anyway does not produce much in the way of net benefits, even if it is "good" research. Then we have the practical consideration that government is often just not very good at carrying out general mandates through specific actions. Even well-justified programs need to be run efficiently. If not, their demise is no great loss either.

As the world's economies grow, the aggregate value of science is increasing because any given research result has more uses. As the amount of research grows worldwide, the number of "free riders" on the results of nonexcludable research is increasing. Together, these two trends increase the justification for federal funding, while at the same time reducing the incentive for private industry to perform fundamental research (which also shifts more of the burden for fundamental research to the government).

What does this analysis of government's role tell us about the likely effects of downsizing science? If we could selectively target the least efficient subsidy programs for elimination, then downsizing might not be so harmful. Unfortunately, science budgets are not established in that way, nor is there any central authority that systematically reviews R&D programs with efficiency as a key consideration. As so many programs have been established with loose justification, it is unrealistic to believe that downsizing would occur on the basis of careful program evaluation. The process of allocating federal funds for science and technology therefore needs to be made more coherent, systematic, and comprehensive. One reasonable approach to this problem is the analysis by a committee of the National Academy of Sciences, *Allocating Federal Funds for Science and Technology*, published in 1995 and updated annually since then. This approach is discussed in the final chapter of this volume.

As we examine the impact of reduced federal spending on specific sectors (federally performed research, universities, and industry) in later chapters, we need to keep in mind the privatization alternative. That is, maybe the nation can survive some budget cuts without too much harm if the institutional setting allows an offsetting private response. Indeed, the private sector's share of total R&D funding has been growing since the 1980s. To what extent can policy capitalize on this trend to counterbalance a reduced federal role? Making it easier for the private sector to pick up some of the research that government no longer wishes to fund will mitigate the losses of research benefits.

6
Just Another Downturn?

How DEEP ARE THE ROOTS of America's inclination to downsize federal spending on science? How strongly will public opinion support science when the discretionary budget comes under baby-boom retirement pressures? The problem is this: if current budget downsizing is only a transitory political lurch in response to budget pressures that could conceivably ease in five years or so, then science funding will make a comeback, and the federal money will flow generously, as in years past. If, however, we are seeing a break in the long upward trend of science's prominence in society, with causes that are deeper and more complex than cycles of federal funding, then there is a much more serious and long-lasting problem. We cannot give definitive answers here, only point out some bad omens.

Worrisome Signs

What worrisome signs, the reader might wonder, could possibly overshadow the enormous evidence of technological progress that permeates our daily lives? All economic trends point to continued technology-inspired growth in the industrial world. Indeed, technical progress seems to be the very order of things, going back centuries. Clearly, this chapter argues a somewhat contrarian point of view.

First of all, consider the views of scientists, many of whom say the situation is terrible. Nobel laureate Leon Lederman has lamented the "end of the frontier." Newly minted Ph.D. scientists complain that they cannot find jobs. Often scientists' calls for more funding are called "whining," and sometimes justifiably so. While more science funding may be good for the nation, it is also good for the scientists who recommend it, thus creating doubt about their ability to make a disinterested case. Most scientists, being intelligent and well educated, already earn comfortable livings. They want *more*? The problem here is that even the most reasoned and objective statements that science should get more funds can be read as self-serving.

Whatever the degree of self-interest involved, the complaints from the scientific community are based on real experiences. Consider, for example, this passage from the widely circulated E-mail message[1] from Alan Hale, the codiscoverer of the Hale-Bopp Comet:

> Like I'm sure is true for many of you, I was inspired by the scientific discoveries and events taking place during my childhood to pursue a career in science only to find, after completing the rigors of undergraduate and graduate school, that the opportunities for us to have a career in science are limited at best and are which I usually describe as "abysmal." ... My personal feeling is that, unless there are some pretty drastic changes in the way that our society approaches science and treats those of us who have devoted our lives to making some of our own contributions, there is no way that I can, with a clear conscience, encourage present-day students to pursue a career in science.

Hale's views are widely held and reflect the experience of many other frustrated job seekers. He received thousands of supportive responses from other scientists.

While the unemployment rates for science appear to be in line with those of other professional occupations, they cover the entire range of scientists and engineers who have been employed for years. The employment problems mainly afflict new Ph.D.s, for whom new jobs are relatively scarce.[2] This weak labor market is a direct consequence of recent funding cuts for science after years of growth, during which the production of doctorates in science continued to grow. If views like Hale's permeate the ranks of bright students in high school, then the pool from which tomorrow's top scientists must be drawn will surely diminish.

Attitudes toward Science

Hale refers to the attitudes of society toward science. What are these attitudes? What degree of scientific knowledge shapes these attitudes? According to a survey performed by the Chicago Academy of Sciences, 40 percent of U.S. adults say they have "a great deal of confidence"[3] in the leadership of the scientific community. While 40 percent hardly seems a ringing endorsement, it is the second highest figure for any institution, exceeded slightly only by medicine, ahead of the Supreme Court and education, and far ahead of the press and television. When asked about benefits and risks of "scientific research," an estimated 72 percent of Americans believe benefits exceed risks, with only 13 percent taking an opposing view.[4]

But when asked about specific technologies, respondents are much less favorably disposed to science. They are only slightly more positive than negative about genetic engineering and about evenly divided on nuclear power. Even the once quite popular space program (which unlike nuclear power holds no risks for the general population) now gets no better than a neutral response.

One might argue that nuclear power is in fact a problematic technology about which reasonable and informed people might well hold negative attitudes. Nevertheless, the long-standing controversy over nuclear power seems to have poisoned public opinion against technologies that are in fact only loosely related to things nuclear but that sound dangerous to some. The irradiation of food, for example, is safe, does not make food radioactive, is approved by the government for use, and is potentially highly beneficial in the prevention of spoilage and bacterial contamination. The only barrier to its widespread use is the food industry's belief that consumers would be afraid of irradiated food or else that "activists" would cause trouble. Distrust of genetic research also interferes with new technologies. Press warnings about Calgene's "Flavr Savr" tomato caused the price of the company's stock to fall substantially, and the company temporarily took the product off the market. European consumers, stirred up by activists and European farm politicians, resist U.S. food products that have anything to do with genetic engineering.

Ignorance of Science

In fact, the public's understanding of science is grossly deficient. Whatever the degree of approval for science, it does not appear to be based on very much knowledge. Consider the surveys of public understanding that the Chicago Academy of Sciences conducts biennially. In the most recent survey, only 44 percent of the respondents said that electrons are smaller than atoms, and only 48 percent said that the earliest humans did not live at the same time as the dinosaurs. As these were true and false questions, a 50 percent score would be achievable by flipping a coin. Only 47 percent believed that it takes a year for the earth to orbit the sun, but this question was exceptionally difficult, as *three* choices (a day, a month, a year) were given.[5] Given these hopeless results for the simplest questions imaginable, how could the public understand biotechnology or other complex scientific matters in the policy realm? One can debate how scientific illiteracy affects people's willingness to support science, but clearly a lot of blind faith must be involved.

Because not many people study science in any depth, widespread ignorance is perhaps understandable. But even those exposed to formal instruction in science, U.S. science students in grades four, eight, and twelve, compare unfavorably with many of their foreign counterparts. In a survey of a half million students worldwide,[6] U.S. students scored below average in math and above average in science (although still significantly below students in such countries as Japan, Korea, Hungary, and the Czech Republic). No doubt our very best students are competitive with the best elsewhere, but in general we are not producing high school graduates whose scientific knowledge is very impressive.

The problem is not that today's scientific literacy is any worse than it was during the years of huge increases in public funding of science. Rather, the question is whether public support for science will collapse under the increasing pressure of tightening federal budgets for discretionary spending. As NSF Director Neal Lane puts it, "In an atmosphere where 'almost no one understands science and technology,' you cannot expect the public to measure the subtle and complex potential of R&D funding against other government expenditures with more direct, tangible results. Under these circumstances, science and technology and the long-term sustainability of the nation could be the losers."[7]

Science and Higher Education

In sharp contrast to the dismal level of public understanding of science, the best U.S. universities are centers of excellence. The science and engineering undergraduate and graduate programs draw tens of thousands of students from all around the world. But the small number of science graduates in relation to the population is insufficient to bring up the average level of scientific literacy. The nation's academic excellence is more in line with the "two cultures" thesis put forth by C. P. Snow, the idea that scientists are a class apart, misunderstood and unappreciated even by the smartest nonscientists.

Perhaps all that we need for public support (backed by public dollars) is enthusiasm for what science contributes to basic human needs. Clearly the strongest support goes to medical research, which gets nearly half of federal funds for basic research. This attitude is reflected in a national survey showing that 79 percent of adults think the government is spending too little on improving health care, a level of priority just under that for education and far ahead of defense, poverty, exploration of space, and scientific research.[8] It is no exaggeration to say that occasionally the National Institutes of Health have been given more money for research on some diseases than they can spend efficiently.

If the public approves broadly of science and thinks the federal government should support it, is this approval showing any signs of eroding? Some writers argue that this is indeed the case, and a number of books decrying the decline of appreciation for science—and even the decline of science itself—have been published this decade. The titles alone of these books set a foreboding tone: *Science under Siege*,[9] *Science on Trial*,[10] *Frontiers of Illusion*[11] (referring to science itself), *Higher Superstition*[12] (about critics of science), *The Flight from Science and Reason*,[13] and, the worst sounding of all, *The End of Science*.[14]

Various recent books on the decline of higher education also have discouraging things to say about the state of science and research. Charles Sykes, for example, in *ProfScam: Professors and the Demise of Higher Education*,[15] directs some stinging blasts against academic research, particularly in the social sciences. Lynne Cheney, in *Telling the Truth*,[16] documents attacks on science in academia by professors whose actual knowledge of science appears to be nil.

Cheney also describes the permanent exhibition at the National Museum of American History entitled "Science in American Life," which portrays science as a leading source of misery in modern American life. "One might well conclude that in the last 125 years, the main accomplishments of science have been the bomb, birth control pills, pesticides, and the ozone hole," she writes. Advances such as penicillin are tucked away in a corner to make room for conspicuous displays of Hiroshima wreckage.[17]

Marcia Angell, in *Science on Trial*, pulls no punches: "The United States is in the midst of a ground swell of anti-science feeling."[18] She says science is now seen by "humanists" as just another discourse constructed by its practitioners, no more objective a source of truth than anything else. Feminists attack science as too male. Afrocentrists attack science as purposely ignoring African discoveries. Antiscience environmentalists flay science for its hubris. To them, the technological imperative has led to the destructive and self-defeating assumption of domination over nature.

Since the 1960s, we have seen how once recondite ideologies have become commonplace in higher education and have wrought many changes that were largely unforeseen. Are we moving toward the time when science departments will lose funding and influence because of academic and "activist" attacks that are now largely ignored and discounted?

A Negative View of Scientists

Beyond the trade books on the decline of science, consider the common portrayal in public entertainment media of the scientist as nerd, villain, or crackpot. The hit 1996 movie *Independence Day* (seen by far more people than the total readership of the books listed above) featured a weird and repulsive government scientist who had wasted billions of taxpayer dollars in his lavish secret laboratory. Devoid of common sense, he had spent decades examining an extraterrestrial flying saucer without figuring out how to make it fly. This turned out to be a simple trick that a savvy environmentalist (!) discovered instantly.

Maybe laughing at Hollywood's version of scientists is just harmless fun. But genuinely costly to public welfare is the rise of "junk science," sloppy or intentionally biased "studies" that cause useful products to be banned and keep others off the market because their makers fear costly legal actions. About fifteen years ago, for example, the media used one small study of babies born of cocaine-addicted mothers to assert that these children would be handicapped for life. In fact, there was no proof that "crack babies" are likely to do worse in later life than others, but the scare thrived because it advanced certain political agendas. Michael Fumento, in *Science under Siege*,[19] documents the "junk science" of the Alar controversy, dioxin, aspects of cancer testing, video display terminals, and other scares. Many, with limited understanding of science, may confuse junk science with real science. Then, when the junk science is discredited, they may become skeptical about the value of the real thing.

Some have suggested that science is in fact declining in its ability to discover new and useful truths. John Horgan's 1996 book *The End of Science* examined the thesis that many of the big questions of science have already been answered and that the answers to those that remain are unlikely to be as significant. He sums it up this way:

> *If one believes in science*, one must accept the possibility—even the probability—that the great era of scientific discovery is over. By *science* I mean not applied science, but science at its purest and grandest, the primordial human quest to understand the universe and our place in it. Further research may yield no more great revelations or revolutions, but only incremental, diminishing returns.[20]

He does not, however, present a sustained technical argument but rather reports his interviews of scientists in a variety of fields who take every conceivable position on the question. That many of these scientists are portrayed as strange and inarticulate, or as monomaniacal megalomaniacs, leaves the reader without much confidence in their opinions. Nevertheless, his thesis may be relevant to some specific fields of science, with implications for the rate of return on research.

Conclusion

The trend does not look very favorable for the future of public support of science. But admittedly this chapter has only skimmed the surface of the future role of science in society. Indeed, one can find books that are optimistic about the future of science and technology (although none that I have seen contend with the problem of paying for the research required to generate that technology). My own interpretation is that the public image of science has slipped over the years, not precipitously but enough to have dampened the enthusiasm for large-scale scientific endeavors and enough to make science vulnerable during tight budget times. This trend will probably continue. Science funding will never be an issue that will win votes for its proponents, because it does not provide a benefit welcomed by a large number of voters, like national parks, school subsidies, student loans, medical care, and tax breaks. Therefore, the individual members of Congress and of the executive branch will need to understand the importance of funding science and do the right thing, regardless of public pressure or the lack of it.[21]

7
The Federal Laboratories—First, Decide on a Mission

THE NEXT THREE CHAPTERS focus on the likely consequences of downsizing for the three principal performers of research—federal laboratories, universities, and industry. The key questions are how federal budget cuts will affect these institutions and their research and what adjustments to new circumstances must be made.

In each of the three cases, the private sector will inevitably assume a greater role as federal funding shrinks. The private sector will not take over everything, of course; there is bound to be less research in total than there otherwise would have been. The question is, How much less? Moreover, these funding changes will have various unplanned consequences for how the institutions of science operate, some of these consequences good, others not so good. Again, we need to think of how to make this change as efficient as possible.

"Privatization" has undeniable benefits. Indeed, a reasoned approach to privatization may help adjust to the downsizing of federal support for science, just as it has in other areas of government. Throughout the world, a keener appreciation of the limits of government has brought a wave of privatization covering numerous government activities, from energy production and telecommunications to prisons and trash collection. Privatization not only helps government perform its legitimate functions more efficiently but also ends governmental responsibility for certain programs, some long accepted as traditionally public that should never have been started in the first place. Can the rationale for privatization be usefully applied to science? That was the object of chapter 5, which set up criteria to delineate the role of government in funding research and development and which we now attempt to apply.

Downsizing Federal Laboratories

This chapter covers research that is both funded and performed by the federal government. The federal government carries out around 10 percent of the nation's research (which it also funds). In 1995, $16.7 billion of the $72 billion in federal funding went for "intramural" research, that is, research in government-owned government-operated labs. An additional $5.8 billion was spent at federally funded R&D centers (FFRDCs), which are government-owned facilities but which are operated by contractors, some of which are universities. Altogether, this research carried out in more than 700 federal laboratories cost $22.5 billion in fiscal year 1995 and constituted about a third of federal funding. The federal labs are a somewhat larger portion, about 40 percent, of the "federal science and technology (FS&T)" budget (the concept that removes testing and evaluation activities, mainly in the Department of Defense, from the definition of R&D).[1] Figure 7–1 shows the inflation-adjusted trend in R&D performed by the federal government. The peak year was 1990, from which there has been a 12 percent decline through 1996.

Their size alone invites the attention of the budget cutters to the federal labs. Indeed, they have become a favorite target for budget cuts, with proposals for downsizing being hurled at them from every direction. Deficit hawks focus on the labs' substantial cost, their only issue being how much to cut. Proponents of further reductions in defense spending, accordingly, would restrict the labs' missions and budgets radically. Advocates of free markets would force the labs to live or die by their ability to sell their services to industry or would perhaps just sell the labs themselves to the highest bidder. All these critics believe that whatever the labs do in the end, they should do far more efficiently.

The primary proponents of the labs are the congressional delegations representing states where the labs reside and the federal agencies that house the labs and for which the labs perform re-

THE FEDERAL LABORATORIES

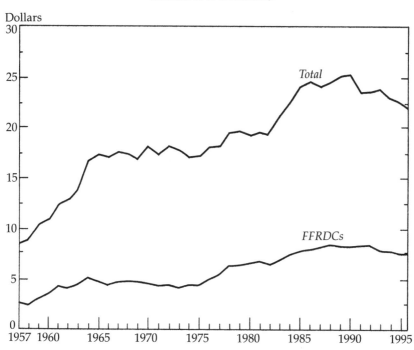

FIGURE 7–1
R&D PERFORMED BY THE FEDERAL GOVERNMENT, 1957–1996
(billions of 1992 dollars)

SOURCE: National Science Foundation.

search. The labs also try to defend themselves by publicizing the importance and usefulness of their work. In addition, some disinterested observers argue that the federal laboratories perform high-quality research, that their management flaws can be cured, and that their missions can be rationalized. One might also list as a "defender" of the status quo the bureaucratic and legislative inertia that obstructs substantive change.

Few would deny that the federal laboratories produce good research. They house enormous scientific and engineering talent. The long tradition of successful, if expensive, defense work is one exemplar. NASA labs helped send men to the moon and back. The National Institutes of Health perform leading-edge research that the public adores. Some of the most advanced work on supercomputers is being done at Sandia, part of the Energy Department's system, in New Mexico. Other labs, like Los Alamos

65

in New Mexico, possess big, one-of-a-kind equipment that is scientifically useful and would be extremely expensive to replace. The Lincoln Labs at MIT perform advanced research for the Air Force. An adequately representative list of the laboratories' successes would be too long to present here. The point is that the labs offer a unique amalgam of scientific talent and leading-edge facilities, with tremendous potential for producing benefits typical of good research.

But most of the laboratories toil under two burdensome handicaps: the importance of their missions has declined, and they are often badly managed by their parent agencies (with an assist by Congress). The logical solutions—get a mission and improve the management—may be beyond the capacity of the institutions as they now exist. Thus, extreme downsizing or outright elimination emerges as an alternative worth study.

What Is the Mission? Federally operated laboratories began their rapid growth during World War II. Everyone knows the story of how some of the nation's best scientists and engineers assembled on a remote desert plateau at Los Alamos and built an atomic bomb in less than three years. After the war, federal labs took on the mission of designing, testing, and building new nuclear weapons. Other installations specialized in weapons design, medical research, and space exploration, all of which are missions in support of well-established national goals. Research on energy production and conservation grew rapidly in the 1970s but has been undercut by the docility of energy markets in recent years.

Federal spending for defense, space, and various forms of basic research can be justified with reference to the criteria developed earlier, if one accepts the missions of the sponsoring agencies. Certainly national defense is a fundamental federal role, and if R&D is needed to accomplish that role, then so be it. Whether the federal government should also perform the research, not just fund it, is another matter. National security and secrecy arguments were paramount in much of this debate, and so the federal labs were assigned to do the research. The federal labs, however, enjoy certain practical advantages over private research companies (related to the ease with which they can be funded compared with the cumbersome system of federal procurement) that raise concerns about unfair competition.

Another issue is just how broad the labs' mission should be. Should they become general purpose research labs, performing research wherever it appears useful? NASA, for example, performs aeronautical research that has contributed greatly to the success of the U.S. civilian aircraft industry. Some of this research is a spillover from its space research, and some is part of an independent mission. But where to draw the line? If aeronautic R&D, why not automotive, farm implement, textile, machine tool, and steel-making research, too? Again we come back to the criteria developed in chapter 5: is the research widely beneficial, can the government do it efficiently, and does it displace research that would have been done anyway? Clearly, we can find examples of federal laboratory research that violate each of these criteria.

None of these concerns is new. Countless studies, reviews, and commissions have studied the labs and made recommendations, the great majority of which have never been implemented. Two reports are particularly noteworthy: the 1983 report by the Federal Laboratory Review Panel chaired by David Packard, prepared for George Keyworth, the presidential science adviser;[2] and the 1995 report *Alternative Futures for the Department of Energy Laboratories*, produced by a task force chaired by William Galvin.[3] Few of the Packard report recommendations were ever enacted, and the Galvin report also appears to be fading into oblivion. The rocklike resistance of the labs' parent agencies to substantial change suggests that the budget may be the only tool available to policy makers who want reform.

Is Privatization a Viable Option? If it were in the public interest to have more energy research performed, and there were no national laboratories, how would we arrange it? First would be a presumption against federal management and control, even though the product of the laboratories would largely be a product for the public. This assumption derives from the basic tenet of the theory of public choice, namely, that bureaucratic control will essentially set its own agenda and seek to achieve the goals of the bureaucratic managers. Foremost among these goals are the survival and enlargement of the bureau, goals that conflict with efficiency. According to the criteria developed in chapter 5, we would expect the bureau to produce some amount of good research to please its congressional sponsors and the public, but we would also expect that research to displace private efforts in the interest of expansion and the bureau to accumulate layers of administrative apparatus that inflate costs.

A better approach is to let the private sector produce the public good. When the army needs a million T-shirts, its best option may be the nearest Wal-Mart. There is no reason to operate federal T-shirt factories or even to contract with a regular supplier, although this used to be the norm. Indeed, private laboratories perform much of the federal research conducted in support of agency missions. Nevertheless, the federal laboratories already exist. Since they possess billions of dollars worth of unique, productive equipment and employ thousands of skilled scientists and engineers, it seems reasonable to explore ways of using this vast resource. The question is, How can we set up the property rights and incentive structure most likely to lead to an efficient process? As we think beyond the current form of government ownership, the key questions would be what to privatize and what to reorganize under a more efficient structure.

Commonly, privatizing a public enterprise entails setting it up in a corporate form and selling stock to the public or selling it in total to the highest bidder, often to a large firm already in the industry. Privatization has proceeded with government-owned utility companies and heavy industry in many countries in this way. To make an informed bid on a public enterprise, prospective owners need to know the conditions under which the privatized business will sell its product and what the demand for the product is likely to be. Will the private firm be a regulated monopoly, like telephone companies used to be? Or will it compete for customers, like a privatized manufacturing company? If the government has always been the primary customer for the services of the enterprise, will it continue to be?

This last question suggests that simply auctioning off the assets of the federal laboratories would not work well. Given the uncertainties of future federal policy and the inability of the federal government to make enduring and credible budgetary commitments much beyond the one-year budget cycle, persuasive assurances that future federal funding for research could be performed by the privatized federal labs appear highly unlikely. In addition, the private sector demand for their services would probably be well under half the labs' total research capacity.

The market value of the federal laboratories would thus be quite low, probably only a tiny fraction of their replacement cost. The point here is not that the government should recover a large sum of money by selling the labs but that their economic value be as high as possible in any new configuration. One example of outright privatization is Harwell Laboratory in Great Britain. Once roughly equivalent to Argonne National Laboratory in the United States, it is now much smaller and less capable, a technical service arm of British industry. Thus, straight-out privatization does not appear to be a reasonable solution, because the broad social benefits from research may well be lost in the process.

Other Consequences of Downsizing "Big Science." Another way to look at the prospective effects of declining federal funding is how "big science" would fare—that is, would the capacity of the labs to perform the occasional gigantic R&D projects that the government takes on be retained? The Manhattan Project was the first such effort; the Apollo program was another. A recent example is the superconducting supercollider, on which billions were spent before the project was canceled. The human genome project and the proposed space station are currently among the biggest of big science.

Quite clearly, big science will be exceedingly hard to fit into downsized federal budgets. NASA's space station is under constant attack, and its budget and hence its mission have been continually reduced. Industry has not shown much interest in manufacturing in zero gravity. The Defense Department has little interest, seeing no particular military benefits to be gained from the space station. The foreign policy rationale is declining, as the Russians are unable to afford their share of the costs. As far as basic research goes, scientists say that cheaper alternatives would do as well. The space station continues as a $2.1 billion item in the FY 1998 budget, but as it is downsized, so too are the justifications for its existence. Journalist Daniel Greenberg commented that "the surviving [mission] is that the space station is being built to acquire experience in building a space station."[4]

The space station is the most recent example of the difficulty inherent in big science. Such problems do not usually result from any lack of technical ability to build whatever is proposed. More often, the justification fades before the project can be completed, or later legislators fail to accept the earlier justification. Sometimes the justifications are inflated or couched in terms of world "leadership" that dissolve on close examination. Inevitably, it seems, the initial cost estimates are far too low.

Adjustments to downsizing, though, are possible. When pressed, scientists can often substitute ingenuity for financial resources and devise smaller-scale projects that are as good as the big-equipment approach. While Livermore's National Ignition Facility is big, it is a substitute for even bigger science, namely, nuclear tests. At Los Alamos, a new Cray supercomputer, capable of 4 trillion calculations per second, will attempt to simulate the dynamics of a nuclear explosion, as well as other complex natural phenomena.[5] In space science, unmanned vehicles have performed most (although not all) of the tasks once envisioned for the much more expensive manned vehicles. "Benchtop" laser-based alternatives to traditional huge particle accelerators are being investigated, and while they are not close to replacing the big machines, they may prove useful in medical applications.[6]

If the downsizing of the federal laboratories continues, it will require a coherent plan, a clear statement of what the central scientific missions are, which labs should carry them out, and how much money will be available to do it. Laboratory directors could then decide which programs to cancel and which to concentrate on. What we have now, however, is a salami-slicing approach, whereby a little is cut from everything each year so that as many programs as possible can remain in existence pending future budget stabilization. This system perpetuates inferior programs at the expense of the best ones. Moreover, gradual downsizing of research units is sure to damage morale and productivity, as opportunities for advancement shrink and hiring of new scientists and engineers is limited.

"Mission Lurch"

The issues surrounding the Department of Energy's national laboratories exemplify the problems of downsizing federal laboratories. DOE controls a network of twenty-six laboratories, each of which is operated under contract by a university or a company. The budget is around $8 billion a year, and more than 50,000 are employed.

These labs are the chief candidates for substantial downsizing, and indeed the aggregate DOE laboratory budget is already on the way down. Their weapons-related missions came under new scrutiny after the collapse of the Soviet Union in 1989 and with the ratification of the Comprehensive Test Ban Treaty, which forbade the testing of nuclear weapons. With these momentous events, the federal demand for additional weapons design, production, and testing decreased considerably from what it had been.[7] Indeed, some writers would redefine the labs' missions into near oblivion. Some assert that the weapons-building mission is obsolete, since (such writers confidently assure us) there is no longer any nuclear threat from Russia or anywhere else in the world.

The mainstream view, however, is that a core of weapons-related activities should remain, along with the responsibility for cleaning up nuclear waste facilities. The three weapons labs, Los Alamos, Sandia, and Livermore, have embraced "stockpile stewardship" as their core mission. To perform this mission better, DOE is building the new National Ignition Facility at Livermore, a nearly $1 billion new unit designed to research the safety and effectiveness of nuclear weapon stockpiles. New national security missions—for example, developing technologies to counter terrorism—are under development. Initiatives such as Sandia's "intelligent systems and robotics" veer away from defense and into improving productivity in manufacturing. While this strays beyond traditional laboratory missions, it may be salable on the basis of broad social benefits unlikely to occur naturally in the private sector.

The Energy Department also claims a role in energy research—alternative fuels, energy conservation, and so on—for its labs. This mission began in 1977 when DOE was established, at which time it was argued that energy sources were dwindling dangerously. Time has proved this prediction wrong, but still DOE research continues on nuclear power, fossil fuel exploration and extraction techniques, and other technologies for private companies "in which the market clearly has no interest."[8]

Given the legitimacy of their core activities and the level of political support they enjoy, eliminating the federal labs is highly unlikely. The practical challenge is to define their current and future missions and then to determine the right size and organizational structure to do the job. Although the labs have all proposed missions for themselves, that is not the same as having a mission firmly established and accepted by Congress and by the public. This book makes no attempt to assign scientific missions. The point is that *some* clear mission must be defined and supported at the highest levels of government before efficient downsizing is possible.

Models of Inefficiency. DOE labs have been widely criticized for inefficiency, the problems stemming largely from overcontrol by Congress and by the department. The labs answer to the Energy Department not only for their basic mission and their programs but also for every detail of their operation. Every nickel and every management decision must be scrupulously accounted for, and the records are subject to numerous audits. Procedures for meeting safety rules are specified in minute detail. Travel requests, even those of top laboratory officials, need several approvals by DOE employees. Cumbersome rules for partnering with the private sector smother many good projects. On top of all this is another layer of direction and occasional micromanagement emanating from Congress.

The Packard group noted the numerous detailed directions imposed on the labs' work, while at the same time the overall missions were ill defined. Even though the labs were subject to countless audits and other niggling interference, the review mechanism focused on evaluation of proposed work rather than on actual performance.

More than a decade later, the Galvin report used similar words to describe the same problems. It cited "increasing overhead cost, poor morale and gross inefficiencies as a result of overly prescriptive Congressional management and excessive oversight by the [Energy] Department."[9]

In congressional hearings on the Galvin report, former NSF director Erich Bloch put it as follows: "The problem with the DOE laboratories is not with their intellectual capabilities, nor with their superb technological facilities, but with the entire DOE organization, its management style, its oppressive controls." At these hearings, Galvin defended his recommendations by saying simply, "It certainly can't be any worse than what . . . [we]'ve got now."

Representative Zoe Lofgren (D–Cal.) said: "[We need] a way to pull that dead mass of bureaucracy and paperwork that stifles innovation away from these labs to allow them to move forward. And in my years of government service at every level, I've never seen the government accomplish that unless you change the way you're doing business." Others expressed their frustration in seeing no management improvement over the years, despite the many recommendations calling for it.

The Energy Department's acting secretary testified that the department had actions underway that would cut the red tape. My own queries of high-level lab and department officials indicate that as of late 1997 some, but not much, progress has been made. Apparently, said one official, the red tape is being cut lengthwise. Problems more serious than excess red tape continue to emerge. The contractor managing Brookhaven National Laboratory, for example, was terminated in the wake of local (and hence congressional) outrage over a tritium leak that was first discovered in 1986 but left uncorrected for a decade.[10]

The Galvin task force's solution to efficiency problems was to "corporatize" the energy labs. By this recommendation, the task force meant transferring the assets of the labs to a not-for-profit organization with a high-level governing board, like the Mitre Corporation, which receives about $500 million annually from the DOD. Mitre, the Lincoln Laboratory at the Massachusetts Institute of Technology, and other DOD units work closely with the Pentagon on technical projects related to military planning, communication, and operations. The Mitre staff works directly with DOD, often in the same offices with regular DOD staff. Thus, they gain a clear understanding of what their assignments are and how to tackle them. The amount of work that these organizations can do is subject to strict budgetary ceilings, and they are always pressing right up against these limits, attesting to the heavy demand for their services.

As attractive as privatization is, this particular approach does not seem to be the answer for what ails the energy labs. It works with Mitre because that organization has a very clear mission, and there is a large, steady demand for its specific services by the Pentagon. The Energy Department, in contrast, does not have the focused mission or the operational needs that would keep a Mitre-like organization productively employed.

Moreover, Mitre appears to be just as subject to oversight by its parent agency and by Congress as the energy labs are, so there would be no automatic reduction in paperwork.

If "corporatization" is meant to achieve substantial independence from DOE, the proposal is not likely to be salable to Congress. Weapons and environmental cleanup missions require close oversight. Basic, self-directed research requires less, but the current mood of Congress does not favor writing blank checks to the labs.

Defining the Mission of the Energy Labs. As the Packard group's report noted, mission creep tends to occur when a national need that justified the original mission becomes a lower priority. Laboratories then move into other lines of work to preserve their existence. Neither the Packard nor the Galvin committees, however, considered preserving the laboratories to be a worthy mission. Going beyond the labs, there is, of course, a case to be made for abolishing the Energy Department itself, along with any energy research mission it might have.[11]

The missions of the DOE laboratories fall into these five categories: (1) national security (nuclear weapons and their stewardship); (2) energy production, use, and conservation; (3) environmental cleanup focused on hazardous wastes at DOE sites; (4) fundamental science, mainly particle physics in large-scale facilities; and (5) industrial technologies. The Galvin task force endorsed all these missions except the last one, saying that only in a few instances are the labs' technologies vital to industry and uniquely available at the laboratories. The labs should sell research to companies when it is expedient and mutually beneficial, they said, but this should not become an independent mission.

Do we need federal laboratories to conduct energy research? The nation's mostly free market energy policy has worked well, in contrast to the regulatory snarl that accompanied the energy shocks of the 1970s. In a practical sense, the world is not running out of energy; markets will adjust by allowing higher prices, which will reduce quantity demanded and give incentives for discovering additional supply. Some of the largest oil-producing nations, however, appear to have short time horizons. They are selling off their oil reserves at prices that may be too low, particularly in the face of the huge future demand for oil that will result from rapid development in China and other Asian nations. Moreover, the political outlook for the Middle East and Nigeria remains problematic; war could disrupt supplies. At some point there may well be another sudden shortage of oil, and in any event U.S. dependency on foreign oil is growing.

Thus, the argument goes, we need research that will develop new and cheaper sources of energy and energy conservation. Private industry may not automatically fill the void left by eliminating government energy research because industry takes no account of the prospective social benefits of greater energy independence. Moreover, one can argue that the energy market has come to expect a compensatory government response (sales from oil reserves and price controls, for example), and these rational expectations induce industry to forgo its own precautionary measures such as research in alternative energy supplies.

At this point, the reader will have noticed that the discussion has coasted down the slippery slope of the "market failure" rationale that was warned against earlier. Where do we insert limits to the federal role, and can we be confident that the government would not commit its own failures? Certainly, the experience with alternative fuels research and subsidization should send up warning flares. In any event, the debate over whether energy R&D is necessary underpins this aspect of DOE laboratory downsizing.

Living with Lower Funding

If we apply the three basic criteria for funding of science that were developed in chapter 5, the federal laboratories will need to undergo some change if they are to hold onto their franchise in the face of relative budget austerity. To assure taxpayers that they produce real benefits, their mission must be rationalized and clarified, and their efficiency must be improved. Based on these criteria and this chapter's discussion, the following general options emerge:

1. *Improve DOE's management of the labs within the current structure.* This recommendation is as naïve as Charlie Brown's annual attempt to kick the football that Lucy holds but always yanks away. Improved management by DOE has been proposed many times, with little success. One is left to conclude that only marginal improvement is in the cards. To give credit where it is due, DOE is trying to reduce the number of audits it requires and has in fact reduced nonresearch staffing at the laboratories. It remains to be seen, however, whether such changes are sufficient to bring real efficiency to the management structure. One trap is to reduce ad-

ministrative personnel without reducing paperwork requirements, thereby diverting scientists from research to the tasks left by the departed administrative staff.

Energy Department officials, in response to the criticism by Congress and the Galvin task force, have said that they intend to improve management and reduce overhead costs. One wonders why these steps were not taken earlier or whether the administrative costs, if lowered, will be allowed to creep up again, in keeping with bureaucratic imperatives. As of the summer of 1995, the department asserted that it was well on its way to accomplishing the managerial streamlining needed to cut costs and bureaucracy. By 1997 various improvements had been made, but much more remained to be done.

Among the most needed changes is improving the labs' ability to sell their services to the private sector. While contracted work will never be the labs' main mission, it is clearly a way to proliferate the benefits of their research. When a lab has the unique ability to perform some research that industry needs, there should be a way to strike an agreement as easily as two private firms could, but the CRADA[12] mechanism, despite recent improvements, remains painfully awkward and slow moving.

2. *Eliminate the labs or severely downsize them.* This step is premature without a careful consideration of the mission of the labs. Some of the missions are vital, particularly those related to nuclear weaponry and cleanup, and the nation would pay a heavy price for neglecting them. Some worry that before long there will be hardly any scientists left in America who have worked with actual, live nuclear weapons.

Unplanned downsizing leaves us with something small and inefficient, rather than large and inefficient—not necessarily a great improvement. In fact, the smaller labs might be even less productive, if they shrink by means of getting rid of scientists while keeping the auditors and other overhead costs. The labs have survived so far thanks to infusions of money sufficient to pay for all the unnecessary regulatory burden, plus having dedicated staffs. But when the money is withdrawn and the bureaucratic handicaps stay, the result will be less money left for research. This situation will further erode the morale of the staff and increase the attrition of the better scientists. If left to drift, labs will become dismal backwaters and strong candidates for dissolution.

3. *Privatize.* Typically, privatization is accomplished by auctioning the assets of the government operation. A very substantial part of the laboratories' assets is their franchise on producing research for the government. This arrangement is a guaranteed source of income, and without it the market value of the labs would be minimal. Therefore, it would be impossible to sell the labs without some clear statement of what the government is committed to buy from them. So once again, articulating a mission for the laboratories is essential to any rational downsizing.

4. *Convene a "laboratories closing commission."* Such an action would be premature and unlikely to accomplish much, in my view. Appointing a commission of prominent people to make recommendations dealing with a political hot potato is a time-honored tradition in Washington. Several years ago, for example, Congress created a commission for closing military bases. Its recommendations to Congress were to be accepted or rejected as a package. The object was to insulate Congress and the administration from an intensely political issue and to insulate decisions from pork barrel politics. Although politics did enter to some extent, the commission was generally believed to have brought a large measure of objectivity into the base-closing process.

Commissions are useful devices when a general policy has been settled on and the appointed panel needs only to investigate and recommend how the policy goals can be achieved. Commissions cannot create policy; they can try, but Congress is likely to reject their attempts. With federal laboratories, too much national policy is unclear or not agreed on. Should federal labs, for example, serve as contractors to private industry? Should they get into environmental research, energy conservation, antiterrorism, missile defense, or some other field in a big way? Or should the labs be given more leeway to perform fundamental research of their own choosing? What current missions should be eliminated? To what degree should privatization be considered an option?

Without explicit guidance on these issues, a laboratory closing commission could do little more than recommend closing a few tiny labs and call yet again for greater efficiency. What is really required is a commission that reviews the basic rationale for the labs' research.

5. *Abolish the Energy Department and house the surviving laboratory functions in other agencies.* Arguments to abolish DOE have

been made in a broader context than the national laboratories they manage.[13] Insofar as the laboratories are concerned, abolishing the department might help achieve the goals of better management and a clearer mission. Certainly, no major change, other than gradual downsizing, is likely under the current organization. The department, for example, has in effect completely rejected the Galvin report's recommendations for changes, specifically, to move the weapons production functions to the Defense Department, to move the waste cleanup either to DOD or to the Environmental Protection Agency, and to move the scientific research functions to an independent agency.

Moving the laboratories to a different agency, however, is unlikely to have any great effect unless their new overseers bring to bear some new paradigm of management that does away with the overbearing bureaucracy that is now so widely criticized. Possibly a new agency would bring new ideas to the task, but the theory of public choice suggests that any other agency would be subject to the same incentives as is the Energy Department and hence unlikely to make radical improvements. If it is decided, though, that the nation no longer needs a national energy policy resident in a cabinet department, then a new home and a new mission for the labs would be required.

Conclusion

The federal laboratories will continue to shrink, given the change in the world situation and hence the mission of the labs. Unfortunately, the present system is not well equipped to cope with the problem of downsizing efficiently. A clear agreement on precisely what role the labs should play in the nation's research enterprise is lacking, and pork barrel considerations are likely to warp any rational plan that may emerge. Therefore, a slow downward drift seems the most likely scenario. The budget reductions will often fall in the wrong places, and morale and efficiency will suffer. As the labs shrink, they will have a hard time attracting the best new scientists.

Clearly, then, a viable mission needs to be defined and agreed upon at the highest policy-making levels of government. This mission needs to be compatible with all three of the criteria developed in chapter 5.

What would such a mission embody? First of all, it should focus on broad social benefits that the federal laboratories are uniquely equipped to produce. Mainly, this would cover the strictly mission-directed research in weaponry, public health, and the like. The missions would entail a certain amount of fundamental research, but not a large-scale transition into research that is best performed at universities. Indeed, the Galvin report went farther than this; it recommended shifting much of the labs' basic research to universities and cautioned against jeopardizing university research by "an exaggerated flow of federal basic research funds to laboratories," a warning particularly apt should funding cuts cause the universities to develop excess research capacity.

More appropriately, the labs could perform applied research that nonetheless has broad benefits and which for that reason companies are unlikely to perform. It should be work that is largely a byproduct of mission-oriented research. In this category might fall advanced computing techniques or intelligent robotics or new manufacturing processes and advanced materials. It would be hard to specify such a mission in great detail or far into the future; this should be left to the laboratories subject to high-level review.

A viable mission would include some research sold directly to the private sector, again work that is mainly a spinoff of existing capabilities.[14] To enable this mission to be feasible, the process of selling research needs considerable streamlining. It is more than just making the agreements with private companies easier to reach and carry out. Efficiency also requires restructuring the contracting process in a way that exposes the laboratories to competition with the rest of the U.S. research enterprise. The process of competition would help identify those nonmission programs that deserve to be eliminated, either because they duplicate research available elsewhere or because they are unmarketable.

In conclusion, a "labs-closing commission" would miss the fundamental issue. Rather, we need a commission—or some other mechanism—to set a clear course for the federal laboratory system. Otherwise a unique and invaluable national resource is in danger of being frittered away in the name of economy.

8
The Essential Federal Role in Academic Research

THE FATE OF ACADEMIC RESEARCH is the most important aspect of federal downsizing. If any activity justifies taxpayer support, according to the theory of public goods, fundamental research meets the test: academic research—particularly fundamental research—is more likely to provide broad social benefits than immediately applicable and appropriable commercial ones. For this and other reasons, academic research has come to depend heavily on federal support, for which there is no ready substitute. The process of allocating federal funds to academic research, while far from perfect, has evolved into a tolerably efficient system, such that we can have reasonable confidence that the funds are going to the "best" science. Thus, the withdrawal of federal money is likely to have substantial effects both on academia and on the nation.

The coming restraint on federal funds will be all the more painful because it reverses a long, rapid upward trend (figure 8-1). Until the past few decades, the United States lacked first-class research universities. According to Mowery and Rosenberg, "Before 1940 there . . . [were] few if any areas of scientific research in which U.S. universities or scholars could be described as operating at the scientific frontier."[1] After World War II, the research universities grew rapidly, and their research activities mushroomed over several decades. Federal support for academic research has grown at an average annual rate of more than 10 percent

FIGURE 8–1
R&D Performed at Universities, with Sources of Funds, 1957–1996
(billions of 1992 dollars)

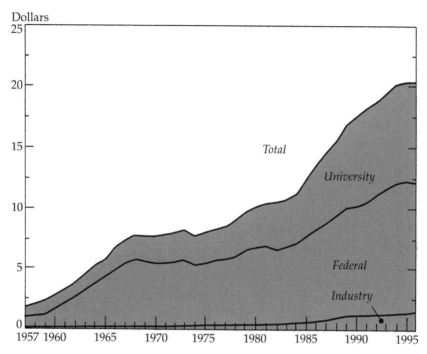

Source: National Science Foundation.

since 1960, adjusted for inflation. Even if we start the series in 1970, to exclude the bonanza of the 1960s, the annual growth rate is still more than 8 percent. Even during the early 1990s federal support grew by more than 5 percent annually.

Universities[2] perform about a tenth of the nation's R&D but half of its basic research. The benefits from basic research are particularly hard to trace or measure. The work of Edwin Mansfield is probably the most widely known attempt to quantify these benefits. In one of his studies,[3] the seventy-six companies he surveyed reported that about 10 percent of the new products and the new processes they use could not have been developed without substantial delay in the absence of academic research. Inasmuch as industry's own R&D expenditures dwarf academic research, and as most of that academic research is not focused on industrial uses,

81

the 10 percent figure is a fairly impressive indication of the industrial usefulness of academic R&D.

The results of university research are in many respects public goods, as they are widely disseminated in journals, conferences, and other modes of communication. The reward structure of the university system gives researchers strong incentives to publish their work. U.S. academic researchers published close to 100,000 articles in scientific journals in 1993, nearly a fourth of the world's scientific articles.[4] Collaboration with industry and federal laboratories, although certainly less open than publication, is another way that knowledge generated at university research labs flows into the world's store of information. Also relevant here is the CHI Research study, cited earlier, that showed that more than half the scientific papers cited on U.S. industrial patents came from research performed at universities.[5] Mansfield's work translates the "spillover benefits" from academic research into economic terms. In one of his latest studies, he estimates a social rate of return from academic research exceeding 20 percent.[6]

In addition, university-based research is an important producer of human capital. In the United States, graduate students are drawn into the research as part of their education. About 90,000 full-time graduate students in science and engineering received their primary financial support from research assistantships in 1993. In particular, the research universities educate the scientists and engineers required by industrial research. In a larger sense, they acquire the skills and knowledge that are used throughout their careers, whether they go into research, education, or some related field such as research management. The net addition to the nation's total stock of human capital would be hard to measure precisely, but it is undoubtedly very substantial. Paul Romer, in weighing alternatives for cutting federal research support, writes, "I would fight hardest to preserve the training of the next generation of scientists and engineers. My reading of the historical evidence suggests that this kind of support for human capital has had a very important long-run influence on our economy."[7]

The Washington Connection

Universities performed 12 percent of the nation's R&D in 1996, and more than half its basic research.[8] The academic share of the nation's research has steadily increased for several decades. Of the $22.4 billion of academic research in 1996, two-thirds went for basic research. Of the nation's doctoral scientists and engineers who report research as their primary or secondary work responsibility, more than a third—150,000 of them—work at universities. Academic research is concentrated in relatively few of the nation's 3,600 institutions. The top 200 research universities account for more than 95 percent of the R&D expenditures in 1993. The top 10 institutions spent 17 percent of the funds.

University research receives substantial support—60 percent of its funding—from the federal government, making it extremely vulnerable to reductions in the federal budget. Some universities are particularly dependent on federal funds: Johns Hopkins gets 89 percent of its research funds from the government; Stanford, 86 percent;[9] Harvard, 74 percent; and Pennsylvania, 74 percent. Federal research dollars are a major source of many universities' total revenues, anywhere from 10 to 20 percent.

Federal support is highly concentrated among the top research universities but not inordinately so when considered as a fraction of their total research spending. The top twenty institutions performed 31 percent of the university research and received 36 percent of the federal funds. The federal government supported 69 percent of their research. Institutions not in the top hundred of research performers still got 56 percent of their research funds from the federal government. Thus, it is likely that cuts in federal funding would affect research at the whole range of institutions. Since research is a far more important budget item at the top twenty institutions than at the less research-intensive ones, however, federal cuts will have a greater impact at research-intensive institutions than at liberal arts colleges that focus on teaching, not research.

Federal priorities for academic research support clearly lie in the area of health. The National Institutes of Health supplies nearly half the federal funds for academic research. The National Science Foundation provides about 16 percent of federal funds for fundamental research in every scientific field as well as engineering. NASA and the Departments of Defense, Energy, and Agriculture provide almost all the rest of the federal funds.

Universities have slightly reduced their dependence on the Treasury for research funds in recent years. The federal share of academic R&D was 67 percent in 1975 and fell to about 60 percent in 1980, where it has remained. Universities have increasingly tapped their own funds (tuition, endowments) to support their research, funding 12 percent of their own research in 1975 and 18 percent in 1996. Industry provides 7 percent of such funding. State and local governments fund about 7 percent, as do other nonprofit organizations such as private foundations.

Industrial funding of university R&D has grown from 3 percent in 1975 to the present 7 percent, something of an increase but still far below the federal contribution and far from any meaningful role as a replacement for federal support. These figures, however, tend to underestimate the connection between university researchers and industry, as they measure direct grants to the university but do not cover consulting fees paid directly to professors.

Industry-university cooperation takes a variety of forms. Wesley Cohen and Richard Florida estimate that 19 percent of research is now carried out in programs that involve links with industry in some fundamental way.[10] One mode of cooperative research is university-industry research centers, which now number more than 1,000 at over 200 campuses, with funding of more than $4 billion in 1994. Around 12,000 faculty and 22,000 doctoral-level researchers participate. But these programs often involve federal sponsorship and receive nearly half their funds from the government, so even in this ostensibly private sector area it appears that federal budget cuts could be painful.

Impacts of Downsizing

At this point, we have no exact estimates of the outlook for funding for academic research. The April 1997 analysis by the American Association for the Advancement of Science indicates that the 1994 to 2002 reduction in federal funds for nondefense research would be close to 14 percent in real terms. As noted in chapter 2, whatever the outcome of current budget plans (always uncertain), the fundamental budget pressures will almost surely force substantial reductions in all parts of the federal research budget.

As a first approximation, the reduction in the amount of academic research might be proportional to whatever the reduction in federal funding might be. Other issues, however, come into play:

- prospects for replacing federal funds with other sources of money
- effects on the scientific labor markets
- effects on the quality of research

Other Funding Sources? Universities have been under financial pressure for some years, so that there are no easy new substitutes for lost federal funding. Rather, there may be some incremental shifts of funding within universities, saving some of the favored research units but in no way making up for the losses that seem to be in store.

Industry support is sometimes viewed as a potential replacement for federal money. That support has been growing, but it still provides just 7 percent of the funding for academic research—and some of that is the direct result of federal encouragement and cost sharing. Industry will not automatically give more money to universities just because federal funds dry up. It needs something in return, including the right to specify what research is to be done, plus property rights to successful research findings. Industry is not likely to finance graduate fellowships in any number. Since education is very much a public good insofar as individual companies are concerned, paying for a fellowship has no payoff for that firm other than marginally increasing the pool of scientists.

One possible avenue for encouraging more industrial support of academic research is through funding coalitions of firms in the same industry. Public power utilities, for example, finance research through the Edison Electric Institute. Other coalitions have formed in semiconductors and microelectronics, with pooled industry research linked to university labs. Such arrangements are most useful when industry can specify the required research and anticipate specific products; they are less well suited to the funding of basic research, where no clear commercial benefits are in prospect. Thus, they appear unlikely to replace very much of the sort of academic research that federal dollars now support.

Some fear that industry funding, to the extent that it does increase, would put too many strings on academic research, diverting researchers from their chosen areas in fundamental research and into specific commercial applications. According to Richard Nelson and Nathan Rosenberg, "Except under special circumstances, we think it ill-advised to try to get university researchers to work on specific practical problems of industry, or on particular product or process development efforts. In general, university researchers are poorly equipped for judging what is likely to be an acceptable solution to a problem and what is not.... [Such] work provides few results that are respected or rewarded in academic circles, unlike research that pushes forward conceptual knowledge in an applied science or engineering discipline."[11]

Apart from direct funding by industry, academic researchers will increasingly seek funding by means of performing patentable research. So far, this has not been a large source of revenue for universities. According to the Association of University Technology Managers, the gross royalties collected by its members in 1994 were roughly 2 percent of their total sponsored research expenditures.[12] From the public policy perspective, royalty seeking may result in research that may be less socially beneficial. By its nature, patented research becomes accessible only for a fee and hence less widely used. The growth of such research may eventually have negative political feedback: taxpayers might at some point object to universities' using public research funds to produce private profits.

Of course, federal funds are not free of strings either, albeit strings of a different sort. In broad terms, the federal budget allocates support to the fields of science deemed appropriate by the political process. More worrisome is the occasional tendency of politics to interfere directly in scientific matters. In 1997, for example, the U.S. Senate went on record as opposed to a report on mammography written by a panel of distinguished scientists,[13] and another quickly convened NIH committee reversed the recommendations. Researchers who failed to find links between the medical problems of Gulf War veterans and their possible exposure to chemical agents were put under pressure to keep looking.[14]

Whatever one's concerns about harmful side effects of industry funding, university researchers will continue to pursue outside support actively, increasing their emphasis on patentable research results as opposed to fundamental research. This trend has persisted for many years, and the coming reduction in federal funding will accelerate such efforts.

Other potential funding sources resemble dry wells. Institutional funds (state and local appropriations, nonfederal grants, tuition, endowment income, and gifts) support around 18 percent of academic research. None of these sources has any particular upward flexibility, and indeed funding by state and local governments has tightened during the 1990s.

Research funding is just one of the ways in which universities receive federal funds. Federal financial aid to students, for example, eventually arrives at universities in the form of tuition payments. In 1992–1993, 45.6 percent of undergraduates and 28.3 percent of doctoral degree students received federal aid.[15] In the current political climate, increasing aid to students has greater appeal than increasing research funding, as evidenced by the 1997 tax legislation that gives tax credits for college tuition and tax preferences to savings accounts intended for tuition payments. Such targeted benefits, and possibly others to follow, increase the aggregate demand for enrollment in universities. This allows universities either to raise tuition fees or to lower the "assessment of need" for students, thus raising students' payments to the university net of financial aid packages. The resulting revenue increases, being fungible, could provide additional funds for research. The general conclusion, though, must be that no other source of funds can replace the amount of federal funds likely to be lost.

Structural Changes. Large reductions in federal funds are certain to change the way universities operate. The "inputs" to research, for example—professors, research assistants, equipment, and buildings—will not all be reduced in the same proportion. There is a natural tendency to protect the people, particularly those with tenure, and let the buildings depreciate. Pressures will mount to use less expensive research equipment, which is not necessarily a bad thing if newer, higher-tech equipment is better and cheaper.

Cost savings may result from new teaching technology (for example, an entire physics course on the World Wide Web or CD-ROM). Of course, such technology would save money only if the university downsizes its teaching staff, which might put even more pressure on research budgets as erstwhile "teachers" try to metamorphose into "researchers."

Another structural change relates to the university system. If we need fewer scientists, then we need fewer university departments that offer doctoral degrees. During the boom years of

1960 to 1985 or so, both the number of graduate students per department and the number of departments grew. The newly created departments were generally not as good as those that had been around a long time. When funding drops, we can expect to see some of the newer departments close down. Just what will happen is hard to predict, as the decision to eliminate departments is intensely political within individual universities and state legislatures, and not necessarily conducive to the efficiency of the university system.

Markets for Scientists and Engineers. As academic research grew rapidly in the postwar years, so too did the production of doctoral scientists. In the mid-1950s, universities were producing around 5,500 doctorates a year; by 1975 this number had grown to nearly 19,000. Clearly this rapid buildup set the stage for an "accelerator effect," whereby any decrease in the demand for research—or even a slowdown—would result in a much larger effect on demand for new Ph.D.s to perform the research.

By the early 1990s, when the funding failed to keep pace with the growing supply of researchers, newly minted Ph.D.s reported difficulty in finding jobs. Going from graduate school into a postdoctoral research position instead of into a tenure-track faculty job became more and more common. Some unsuccessful job seekers would take two or even three postdoctoral appointments in succession. Unemployment rates for young scientists remained well below those for the work force as a whole, but certainly there had been a keenly felt change in the market for scientists. Recall the message of astronomer Alan Hale, quoted in chapter 6, discouraging young people from entering scientific careers because of the lack of good jobs. The proximate causes for the weakening labor market included an economic recession, defense-related R&D spending cutbacks, and the downsizing of some corporate labs. But the overarching reason was that the production of new Ph.D.s had been unsustainably high for some years.

As fewer young scientists are hired, the work force ages. The rapid hiring of faculty during the 1960s and early 1970s brought in many new Ph.D.s and lowered the average age of the research work force. But as that cadre got older and hiring slowed, the professoriate aged markedly. In 1973, the median age of full-time doctoral science and engineering faculty was 40.3 years; by 1993, it had increased to 46.0 years. In 1973, 30.3 percent of the faculty was thirty-five or younger; in 1993, only 11.9 percent was that young.[16] This aging process leveled off starting in about 1990, but new data will probably show that it has resumed. A new phase of constrained hiring would result in still more aging.

To the extent that younger workers are more productive, this aging of the academic work force will diminish the quality of research that is performed. It is generally believed that scientists make their most original contributions during their early years of research, when they are in their twenties, thirties, and forties. There are many exceptions, of course, and the age effect is not easily quantified.

Perhaps older scientists are less receptive to new findings than younger scientists are. According to Nobel Prize–winning physicist Max Planck, "A new scientific truth does not triumph by convincing its opponents and making them see the light, but rather because its opponents eventually die, and a new generation grows up that is familiar with it."[17] According to this view, science is not a continual building process, but an activity that is both creative and destructive. Obsolete theories and paradigms must be junked if progress is to occur: no young scientists, no new paradigms.

According to this interpretation of aging, the output of science would diminish more than simply in proportion to cuts in spending, particularly if an accelerator effect greatly reduces the number of young scientists. It also suggests that nations where science is growing rapidly might get an extra boost in research quality because of the youthfulness of their scientific work forces.

The Quality of Research. Certainly the quantity of research suffers from a lack of funding, as fewer scientists perform research, but the quality of research—the output per scientist—is likely to suffer as well. In addition to the consequences of an aging research faculty, morale and efficiency deteriorate in any static or contracting institution. When chances for personal advancement fade and when it becomes harder to attract support for new ideas, enthusiasm for the research enterprise dwindles.

Downsizing, of course, has haunted the corporate world for the past decade. Recent evidence suggests that quality has suffered, according to a study by the National Research Council funded by the Army Research Institute.[18] The study concludes that "downsizing as a strategy for improvement has proven to be, by and large, a failure." Downsizing, even if accompanied by plans for "reengineering" and "total quality management," often hurts the performance of the surviving workers.

Peer review—the process used most frequently for awarding federal grants to university research—will also suffer from downsizing, with consequences for the quality of research in U.S. universities. Managers of federal research programs solicit confidential reviews of research proposals and depend on them to decide which proposals to fund. Peer review is designed to encourage free and fair competition among ideas and to keep pork barrel politics from corrupting the awards process.

But peer review is costly, not only in dollars but also in time. Research scientists spend large blocks of time writing proposals, and other scientists spend additional time reviewing them. As federal funds decline, success rates of proposals will drop, with the result that the cost of peer review per successful proposal will increase. In time the "market" for proposals will adjust and, because of the reduced probability of being funded, fewer proposals will be submitted. But before this (none too desirable) adjustment takes place, the peer review process will become more onerous and less accurate in assessing proposals. With less research money to go around, reviewers may become less objective, as they see that any grant that they approve will deplete the funds available for others—including themselves—within their own narrowly defined field.

Conclusion

If federal research budgets decline according to the scenario outlined in chapter 2, academic research will pay a sizable price. Given the nature of the benefits from this research, the nation will eventually pay a price as well. Not just the quantity but the quality of the research is likely to fall. To be blunt, reducing funds for scientific research will hit the nation's best universities much harder than it will those in the middle of the pack or at the bottom.

Reflexive advice to universities like "be more efficient" and "get more funds from nonfederal sources" is not particularly helpful. Universities have been under financial pressure long enough to have examined all the angles of fund-raising. Likewise, universities may be close to the limit of how much they wish to tap their general funds to support research. Rapidly rising tuition fees have been widely criticized,[19] and universities have defended themselves with evidence of rapidly rising costs for necessities like security, heating, libraries, and building maintenance. True, the top research universities have the largest endowments, a few running well into the billions of dollars; it will be interesting to see how much those endowments will be tapped to rescue research. Whatever the actual degree of financial pressure, universities are unlikely to replace very much lost federal research support with their own funds.

One reaction that will certainly continue is the effort to earn more money from academic research. This attempt will take various forms: seeking grants from or contracts with industry, establishing more profit-oriented consortia, developing more individual consulting, and so forth. This course of action is dangerous in some ways, for the wide-ranging social benefits of freely available research may be compromised. Another concern is that industry will divert academic research into mundane projects with short-term payoffs but with lower ultimate benefits to society. As universities ratchet up their quest for patent income, the structure of academic research may change in ways difficult to foresee but that will require serious scrutiny. Understanding this process is vital to managing the process of downsizing, because intellectual property rules need to be defined so as to avoid unwanted changes in the structure of university research.

Whatever the dangers it holds, this "privatizing" trend is already well underway. While continued growth in federal support would obviously be more favorable to the social benefits from fundamental research, this is not the real alternative. Rather, the trend toward commercialization of academic research seems clearly preferable to simply letting academic research shrink: better some commercially driven research benefiting only its sponsoring company than no research at all.

9
Improving the Environment for Industrial Research

INDUSTRIAL RESEARCH—R&D performed by private companies—now constitutes 71 percent of the nation's R&D, a share that has fluctuated over the years but never by more than a few percentage points from what it is now.[1] Industry performed an estimated $132.1 million worth of research in 1996, of which industry itself funded 84 percent and the federal government funded the rest. The share of industrial R&D funded by government has gradually decreased over the years, from more than 50 percent in the early 1960s to just 16 percent now (figure 9-1).

In addition to performing most of the nation's research, industry is also the most important source of research funds. Industry funded about 62 percent of the nation's R&D in 1996. Of the $113 billion that industry spent on R&D, nearly 98 percent went for its own research; the rest funded research at universities and other nonprofit organizations. Thus, industry is the largest performer and funder of the nation's R&D. Nonetheless, government funds and policies materially influence the size and the direction of industrial research.

What Price Federal Downsizing?

In assessing the effects of federal downsizing on industrial R&D, we first need to consider two questions: what are the benefits from industrial research, and what is the legitimate role of government

FIGURE 9–1
Federal and Industry Funds for Industrial R&D, 1957–1996
(billions of 1992 dollars)

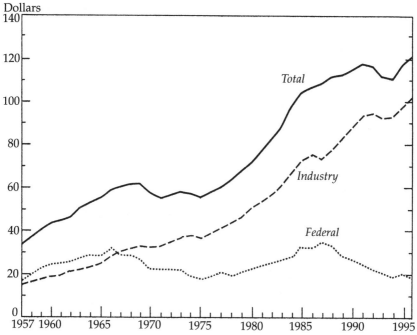

Source: National Science Foundation.

in this regard? Those questions will provide a framework within which we can investigate whether current budgetary plans are likely to reduce the social and economic benefits that industrial R&D is supposed to deliver.

Benefits from Industrial R&D. The economic benefits from industrial research are very substantial and obvious to observers of the national economy. The estimates of benefits discussed in chapter 3 pertained most specifically to industrial research, much more so than to federally performed research. Indeed, most of federal funding of industry goes to buy R&D that federal agencies (mainly the Defense Department) need to carry out their own missions, research that can lead and often has led to private sector economic benefits but is usually not designed for that purpose.

When firms pay the bill, however, their objective is clear: to acquire knowledge and translate it into processes, products, and solutions that create competitive advantage for the enterprise and financial benefits for shareholders. The company is buying knowledge for its own use, not for society at large and certainly not to help its competitors. While broad social benefits can certainly result from industrial research, they play little or no role in the firm's decision making. Firms' decisions on R&D, it seems reasonable to assume, are based on profit-maximizing criteria. Social benefits are largely incidental.

The management of industrial research, however, has received much retrospective criticism. During the past twenty-five years and more, sundry reports have argued that firms make all kinds of mistakes regarding R&D: time horizons are too short, there is not enough focus on manufacturing processes, too little attention is paid to applications of new technology, and other such scolding.[2] From these assertions follow recommendations that chief executive officers wake up (that is, follow that particular report's recommendations) and support for proactive federal policies to stimulate industrial R&D, often by means of subsidies.

Unquestionably, corporate managers make any number of poor investments and bad decisions, but it does not follow that they are incompetent or need corrective direction from government. Rather, firms try to maximize profits in an environment where, because information is costly, there are limits to how much information the firm can profitably collect. If companies did not have *some* failures, it would mean that they were not taking enough risks. Managers gather as much information as is worthwhile, but at some point, additional information costs more than it is worth. Prospective results of research and innovation are particularly uncertain; indeed, such information is unavailable at any price. Moreover, many important events external to the firm are unpredictable regardless of how much a priori information is collected and analyzed. The government is unlikely to be able to assist the R&D investment decisions of firms by supplying additional information or imposing its own judgment about where to direct research funds. Indeed, ill-advised government actions are often costly to industry and a source of instability in the economy.

Government's Role. Nevertheless, government policy needs to pay attention to industrial research. Like any other investment, R&D must be allocated efficiently. Just as the government should create an economic environment that encourages firms to make economically efficient decisions on investment in plant and equipment, pricing, hiring, and other operations, it should try to ensure that optimal R&D decisions are not stifled or distorted by bad federal policy.

This objective requires a well-designed and well-enforced system of patents and intellectual property rights, so that firms have a monetary incentive to pursue new knowledge. Regulatory policy needs to be efficient as well as consistent, so that the introduction of new drugs, for example, is not delayed unnecessarily or precluded by an abrupt change in policy. Tax policy toward R&D should be patterned after optimal policy regarding investment, meaning that the market rate of return should equal the effective rate of return to the firm and that countless possible distortions of incentives are avoided.

As important as these policies are, however, they are beyond the scope of this book because none of them is directly threatened by any of the budget cuts currently planned. The germane policies are federal funding of industrial research and federal funding of other research (especially university-performed research) that is useful to industry. More specifically, we will look at federal funding of research that contributes to agency missions and direct subsidies such as the Advanced Technology Program administered by the Commerce Department. (The tax credit for research and experimentation might also be considered a subsidy, but it is not part of the budget cuts under consideration.) Thus, the "downsizing" issues for industry mainly concern, first, government as a customer for R&D performed by industry and, second, subsidies of one kind or another. The longer-term—and likely more important—issue is industry's potential loss of benefits from downsized academic research.

How Healthy Is Industrial R&D?

As federal funding for R&D declines, industrial research will need to assume a greater responsibility for the nation's technological progress. Is it in a good position to do so?

Funding by Companies. More than 80 percent of industrial R&D

is financed by the companies themselves. The federal government is an important but declining source of funds for industrial research. Company funding of its own R&D is not only much larger than federal funding, but it is also likely to be the source of the greatest economic benefits, since it is chosen by the firms for precisely that purpose.

Over the long term, company-funded industrial R&D has grown reasonably well. From 1960 to 1996 it grew at an average annual rate of 4.3 percent, adjusted for inflation. That pace exceeds the annual growth rate of the economy (3.1 percent over that span) and is on a par with the growth of other forms of business investment.

Industrial R&D behaves much like other forms of business investment, varying in response to business conditions and changes in the outlook for particular research-intensive industries, such as computers, autos, and drugs. Company-funded R&D is actually somewhat less variable over time than industrial investment, as measured by investment in producers' durable equipment. Not surprisingly, the two series are highly correlated.[3]

Company funding of industrial research, while not highly volatile from year to year, is nevertheless somewhat unpredictable and has occasionally slumped. For example, industrial R&D fell sharply in the early 1970s, recovered, surged in the late 1980s and early 1990s, leveled off in the mid-1990s, and now appears to be growing again, rebounding from earlier cost-cutting and reorganizations in corporate R&D structures.[4] Such slumps were neither predicted in advance nor have they been well explained after the fact.

Many reports over the years have decried the perceived inadequacy of industrial R&D.[5] Of course, with any data series that varies from year to year it is easy to pick out multiyear periods when the growth rate falls below its long-term trend or below some previous peak. The growth rate of U.S. industrial research during such slow periods may also fall below that of selected foreign countries, providing grist for findings of "lagging competitiveness." Data on industrial R&D are also revised often, lessening the confidence one might have in the significance of whatever year-to-year changes these data suggest. Thus, long-term trends are better guides to assessing industrial R&D.

The Role of Federal Funds. Such reports, however, usually focus on total industrial R&D, that is, company funding plus federal funding of R&D performed by industry. Not only is the federal component less stable than industry funding, but also it has trended downward for some time. Federal funding rose in the 1960s (led by NASA and DOD), fell in the 1970s, rose again in the 1980s (defense R&D), peaked in 1987, and has been falling ever since. Over the years, federal funding has been a substantial source of variation and decline in industrial R&D. As its share has dropped (from nearly 60 percent in 1960 to under 18 percent in 1996), however, so too has its influence on the total.

Gyrations in federal funding, though, are not the only source of instability that industry has to contend with. Other policies have vacillated over the years and for that reason have lost much of their effectiveness. Here the most obvious example is the tax credit for research and experimentation, which was initiated in 1981 and through 1994 had reduced companies' tax liability by around $24 billion. But during these years the tax credit was changed considerably, has had to be renewed almost annually, was allowed to lapse several times, and helped only if a company was *expanding* its R&D over the prior year. Because of the uncertainty caused by these vacillations, the effect of the tax credit on firms' actual R&D spending was diminished. Firms were understandably reluctant to be lured into large multiyear commitments to research in hopes of collecting on a credit likely to be rescinded without warning. Evaluations of the effect of the tax credit have been somewhat inconclusive, largely for this reason.[6] In addition to these macro effects, governmental inconstancy of effort also affects the success of particular programs. Projects that involve federal-private partnerships lose their appeal to industry, and therefore their chance for success, if it is suspected that government will bail out prematurely.

Whatever the long-term instability in federal funding, the federal budget has clearly entered an extended phase of cuts in spending on mission-oriented research that would be performed by industry. The largest single piece of that reduction is military R&D, which is dropping substantially. According to an analysis by the American Association for the Advancement of Science, defense R&D fell by 2.1 percent in constant dollars from FY 1994 to FY 1997 and is slated to fall by another 17.8 percent from FY 1997 to FY 2002. As military R&D constituted 54 percent of federal R&D in FY 1996 and the preponderance of federal funds to industry, it is clear that defense downsizing is the most important single contributing factor in the reduction of federal involvement in industrial research.[7]

The real issue here is whether this massive cut in defense R&D will eventually degrade the nation's military preparedness, but such questions are beyond the scope of this book. Given the number of potential eruptions around the globe, and given the U.S. reliance on superior military technology rather than on sheer numbers of soldiers, such reductions deserve serious review from the standpoint of long-term military strategy.

Is Industry Abandoning Basic Research? In recent years, industry is believed to have shifted some of its own R&D spending away from basic research in favor of applied research and development. According to Richard Rosenbloom and William Spencer, "Industrial research in the United States has been transformed in the 1990s. Research managers have had to redirect their resources away from fundamental science and pioneering technology toward activities that are more relevant to current product and process development, more likely to produce results that can readily be kept proprietary, and more certain to produce a commercial payoff in the near future."[8] This pattern is certainly evident at many of the largest, most research-intensive companies, such as IBM, AT&T, and Kodak (although it does not mean that high-risk investment in R&D has disappeared from these or other firms).

A related trend is the decentralization of research responsibility and funding support within firms. Less work is being assigned to the large central laboratories, and more is going to labs affiliated with the operating divisions. The object of decentralization is said to be to tie research more closely to the short-term needs of business units; more applied and less basic research will be the result.[9]

Several possible reasons have been given for this shift. First, it may be that a sober analysis of the economic payoff to basic research has come out in favor of more short-term research. Indeed, some have called the era of basic research performed by big labs a "passing fad." Second, with the growth of science and technology throughout the world, firms may see that the results of their basic research inevitably flow to their domestic and overseas rivals. In a parallel reaction, scouting the world for someone else's research results has become relatively more worthwhile (there are now more new discoveries out there than ever before) than pursuing one's own, more costly, fundamental research. Sophisticated new information systems have facilitated the search for other people's discoveries.

The data from the National Science Foundation survey on basic research tell a more ambiguous story. Industry performance of basic research increased by nearly 50 percent in 1991 and then declined in following years. The suspiciously large increase in 1991, however, is believed to be largely the result of changes in the survey.[10] Over a longer span of years, basic research as a share of R&D funded by industry increased from 4.1 percent in 1980, to 5.3 percent in 1990, to 6.5 percent in 1995. Possibly the downsizing of some of the large corporate labs has been counterbalanced by increases at smaller, less visible companies.

To the extent that industry is moving away from basic research—the form of research that generally has the greatest social spillover benefits—a greater burden is placed on the federal budget for the funding of truly "basic" research. Subsidizing industry to perform more basic research holds little promise; industry would accept the money but is unlikely to rival the best universities as a generator of broad social benefits. There is no reason to believe that industry would be more cost effective in performing research than universities. Moreover, industry lacks one crucial ingredient of the recipe for broad social benefits, namely, graduate students. Many have noted that the combination of research and education unique to universities is vital to the contribution that universities make to the nation.

Is Industrial Research Becoming Less Productive? There is some evidence that industrial research is not as productive as it used to be. Bronwyn Hall concluded that the private returns to R&D in U.S. manufacturing declined between the 1960s and the 1980s.[11] That decline may not be evidence of something "going wrong" with industrial R&D but rather the result of an equilibrating process, whereby industries have raised R&D to the point where its rate of return approximates that of other forms of investment.

Patents per dollar of industrial R&D and patents per scientist in industry both fell fairly steadily between 1925 and 1991. While this trend is consistent with declining productivity of research, it may simply reflect a change in the economic meaning of patents over the years.[12] In some fields like software development, for example, the time horizon is often too short to make the patenting process worthwhile.

Other analysis at the industry level suggests problems with productivity. Henderson and Cockburn concluded that "over the past 20 years, the pharmaceutical industry appears to have suffered a dramatic decline in productivity." They raise the possibility that higher costs of research reflect decreasing returns. They speculate that the soaring cost of developing clinical drugs also "reflects both a shift to the treatment of conditions that require more complex clinical trials and increasing regulatory stringency."[13]

Taken together, this and other evidence imply a decline in the rate of return to industrial research, a decrease in the ratio of results to effort, which is to say a reduction in productivity. Such conditions undercut the argument that the private sector can take up the slack left by downsized federal funding. They also suggest that the federal policies pertaining to industrial research need to be reexamined to see whether they are affecting the productivity of industrial research.

Some would argue that recent downsizing has made industrial R&D more efficient, but this assertion needs empirical confirmation. Conceivably, downsizing has caused disruptions that have reduced output per researcher.

More Collaboration. Increasingly, companies are collaborating on research among themselves and with universities. Here the data are somewhat scattered, but the trends seem clear. Cohen, Florida, and Goe, for example, find a substantial increase in industrial funding of university research.[14] Possible reasons for this increase include higher costs of industrial R&D and the fact that the growth of university research has made it easier to find academic expertise in any particular field.

Federal policy has also encouraged collaboration. Since about 1980, federal policy has favored collaboration within industry, industry with university, and government with industry. Specific legislation includes the Bayh-Dole Act of 1980, the Stevenson-

Wydler Technology Innovation Act of 1980, the National Cooperative Research Act of 1984, and the Federal Technology Transfer Act of 1986. Some of these (and other) initiatives simply allow firms more latitude in joint ventures. Others, such as the national Competitiveness Technology Transfer Act of 1989, that allow government-owned contractor-operated laboratories to enter into cooperative R&D agreements are aimed at better use of federal resources. Still other procollaboration policies include funding to support various types of joint ventures or require industry cost sharing of projects funded by government.[15]

These actions grew out of the realization that federally funded research was not being used widely enough and that some federal policies (particularly antitrust regulations) were standing in the way of efficient collaboration among companies. The resulting programs are all essentially efforts to get more results out of the nation's research spending by easing the flow of information and by spreading fixed costs in ways believed efficient.

Reactions of Industry

In a discussion of industry's reactions to federal downsizing, the two main issues are, first, whether industry funds will fill the gap left by reduced federal spending and, second, how badly industry would feel the loss of federal funds that support industrial research.

The Spending Gap. Will industry take up the slack in areas where federal funds are cut? Some advocates of a reduced federal role argue that private sector R&D will increase in an amount sufficient to offset any reduction in federal funding. This scenario appears highly unlikely, though. While it is true that industrial research has recently grown while federal funding has fallen, there is no evidence of a causal relationship between the two.

Industry might be inclined to support research that was formerly federally funded if it produced specific benefits for particular firms. That condition applies to only a small part of federally funded research at universities, however, most of which has general, non-firm-specific benefits. Industry still provides a small proportion of the financial support for academic research, although its share is growing, from 3.9 percent of academic research in 1980, to 6.9 percent in 1990, to an estimated 7.1 percent in 1996.[16] Thus, while we can expect increases in industry funding of university research, it will remain concentrated on specific technologies, sometimes funded by consortia of companies to spread the costs.

Loss of Subsidies to Industry. Would industry miss government subsidies for research? Quite a number of programs exist for the purpose of correcting alleged market failures by selectively targeting money at industrial research or technology development. Some of these programs pay industry to do the research; others conduct the research in federal laboratories and make the results available to industry. The largest such programs and their FY 1997 appropriations follow:

- *Advanced Technology Program* (Commerce Department, $225 million). Makes grants to help companies apply basic research in new technologies. Recipients include such giant firms as Xerox, General Electric, IBM, and Caterpillar, plus many smaller firms.

- *Manufacturing Extension Partnership* (Commerce Department, $95 million). Analogous to the Agriculture Department's venerable Extension Service, the MEP funds the creation and maintenance of dozens of extension centers to assist small and medium-sized businesses using modern manufacturing technologies.

- *Agricultural Research Service* (Agriculture Department, $786 million). Conducts research to increase productivity of land and water resources, improve farm products, and find new uses for those products.

- *Partnership for a New Generation of Vehicles* (multiagency, $240 million). Provides research funds to Chrysler, Ford, and General Motors (but not Toyota, Honda, or other foreign companies with large U.S. manufacturing plants) to help develop a car that is much more fuel efficient than today's models.

- *Aeronautical Research and Technology Activities* (NASA, $888 million). Covers NASA's entire effort to improve aeronautical technology, some in NASA labs, some by assistance to industry.

- *Defense Advanced Research Projects Agency* (Defense Department, $1,111 million). Includes grants to large firms such as Boeing and Texas Instruments to develop dual-use (that is, defense and civilian) technologies.

- *Energy Supply Research and Development* (Energy Depart-

ment, $2,711 million). Covers a wide range of energy research in national labs, universities, and partnerships with industry related to solar and renewable energy, nuclear energy, and fusion.
- *Fossil Energy Research and Development* (Energy Department, $365 million). Covers research performed in national labs, universities, and partnerships with industry related to clean fuels, oil technology, natural gas, and fuel cells.
- *Clean Coal Technology Program* (Energy Department, $12 million). Funds joint public-private demonstration projects to help industry develop cleaner coal-burning technologies.
- *Commercial Space Transportation Office* (Transportation Department, $6 million). Funds R&D to encourage private-sector space transportation.

Altogether these programs absorb an amount somewhere in the low billions of dollars annually for direct funding of industrial research. They are justified as ways to "help remedy the underinvestment in R&D created by the limitations on private appropriability of its economic returns."[17] In other words, the idea that spillover benefits may exist is applied liberally to a wide variety of technologies and federal objectives, in most cases without rigorous evaluation of results. Let us consider one of these programs more closely and see how it fits in with the criteria for federal spending developed in chapter 5.

The Commerce Department's Advanced Technology Program (ATP) is designed to stimulate applied R&D projects by providing matching grants to companies that develop technologies useful across industries. Award recipients must pay at least half their project costs. Since the early 1990s, the program has expended more than $1 billion in matching grants to nearly 300 firms and consortia. The Clinton administration originally proposed annual budgets as high as $750 million for the program, but the ATP has always faced opposition in Congress, with the term *corporate welfare* applied repeatedly. The president's FY 1998 request was for $275 million.

Because the program is so new, and because the rewards to research are long term by their nature, no useful evaluations of the program's actual benefits now exist.[18] When such evaluations do appear, they will of necessity be focused on individual projects rather than on the entire program. Thus, at present the only way we have to assess the program is to evaluate its design, mode of operation, and goals. In effect, based on what we know now, what is it likely to accomplish?

Loren Yager and Rachel Schmidt use this approach in a recent monograph.[19] They apply the criteria of whether a given project would have been funded in the absence of a federal subsidy. If so, then the ATP simply constitutes a transfer payment to the firm, with no effect on spending for R&D. Federal money would merely displace private money.[20] To be cost effective, the program needs to choose projects that firms would not have selected because of their low private payoff but which produce high social benefits. For the program to work right, the firms would have to propose projects whose benefits will accrue largely to others. Firms' obvious incentive, however, is to propose projects that will benefit themselves. As Yager and Schmidt show, the procedures used to select projects are not likely to lead to such altruistic choices.

The authors conclude that the ATP program has had "only limited success" in selecting projects that could not obtain funding within the private sector. Most of the projects would have been done anyway, they conclude. This finding is consistent with a survey conducted by the U.S. General Accounting Office, where half the respondents who did not receive ATP funding were able to continue the research project without federal funds.[21]

The Yager-Schmidt findings imply that funding to the ATP could be reduced without much reduction in the amount of industrial R&D performed. Nothing in this conclusion disparages the quality of the research performed (which is probably quite good). The problem is simply that the program does not deliver many benefits that would not have occurred independently. It would be worthwhile to apply similar evaluation techniques to the whole range of technology programs supported by federal funds.

Other Federal Policies and Funding

The programs discussed above were designed to affect industrial technology directly: they spend federal funds on research performed either in federal labs, universities, or industrial labs. But federal policies with goals unrelated to industrial research are probably more influential in the aggregate. Inasmuch as industrial research is an economically motivated investment, the general economic climate will heavily influence companies' decisions to spend money for R&D.

Federal regulations also affect research. Research in the pharmaceutical industry, for example, is intensely dependent on regulatory issues; a single FDA decision on a particular drug can completely wipe out the return on the investment that developed that drug. Applied more generally, the changes in the perceived probability that the FDA will disapprove a new drug, or the expected number of months that a decision will be in process, will move the rate of return on research up or down. When regulators unduly delay or deny approval of new pharmaceutical products, or require more expensive testing than is needed, they impose costs on consumers and producers; this is the basic idea behind economic analysis of the regulatory process. In addition, bad regulatory decisions effectively reduce the rate of return on investment in R&D and therefore cause a reduction in industrial R&D.

The legal environment also affects industrial research incentives. Marcia Angell, in *Science on Trial*, points to the U.S. tort system, which, she says, "enables lawyers to prey on people's fears, to destroy thriving companies, and, in the breast implant case, even to threaten an entire industry (medical devices) and an important area of medical research (epidemiological studies)."[22] She cites the case where DuPont spent $8 million defending itself for supplying Teflon to the manufacturer of allegedly defective jaw implants,[23] pointing out that if a firm loses just one out of many such cases the costs can be enormous. In another example, Merrell Dow spent over $100 million defending itself against lawsuits alleging that Bendectin causes birth defects.[24]

As large as the measurable legal costs of this tort lottery may be, the hidden costs may be even greater: the ever-increasing chance that a new product may land a firm in court is a clear incentive to reduce research and innovation.[25]

Conclusion

Over the next several years, industrial research will not suffer much from the downsizing of federal science. In the longer term, the problem may be substantial.

Federal funds for industrial research have dwindled over the years, and the loss has seemingly been absorbed without great problem. It remains to be seen just how well the various research subsidy programs will fare in the budget battles to come. These programs will probably maintain enough political support to insulate them from elimination. If they are reduced in proportion to the rest of the R&D budget, the loss to industry would be only a small portion of total industrial research.

Would the economy lose much if subsidies to industrial research were cut back? It is hard to assess the success of the various programs in accomplishing their objectives, given the long-term nature of research results. While they probably produce some good technologies, they are not necessarily cost effective. The question with subsidies comes down to this: subsidies are justified only to the extent that they buy social benefits. Benefits that accrue only to the firm are not legitimate goals for federal subsidies, if one accepts the market-oriented framework for economic policy.

The recommendations of the NAS report, *Allocating Federal Funds to Science and Technology*, bear repeating:

> The government should encourage, but not directly fund, private-sector commercial technology development, with two limited exceptions:
> - Development in pursuit of government missions, such as weapons development and space flight; or
> - Development of new enabling, or broadly applicable, technologies for which government is the only funder available.[26]

Given that government is almost never the "only funder available" for worthwhile technologies, this recommendation would seem to be equivalent to "no subsidies for commercial technology." Indeed, the report voices skepticism that such subsidies would ever be the most efficient use of federal dollars.

The fundamental problems for industry will not come from reducing federal funds for industrial research but rather from the downsizing of academic research, which underpins so much of technological change in industry. As discussed in chapter 8, academic research is vital to much of the applied research and development that take place in industry. Given the subtle and long-term nature of this connection, few bad effects would be noticeable, let alone measurable, until many years from now. This loss is unlikely to be offset by increased funding of university research by industry. It will also be very hard to repair the damage to the research infrastructure, should some future government wish to do so.

If the nation chooses to reduce federal funding of research, its dependence on industrial R&D will increase. Therefore, industrial research must get every chance to succeed. Few would disagree with this. Indeed, every administration claims to be well disposed toward science in general and industrial R&D in particular. The problems arise from the indirect effects of programs and policies with other objectives that collectively hamper industrial research. The other major failing of federal policy is its inconstancy and unpredictability, as evidenced by the cyclicality of funding and by the off-again-on-again tax credit for research and experimentation.

10
International Dimensions of Downsizing

BY ANY CREDIBLE MEASURE, the United States leads the world in scientific research. It has done so for more than fifty years. But if the reductions in federal funding that have been proposed actually occur—or even if past trends continue, for that matter—then this leadership will erode. To be sure, the United States is certain to remain the largest single performer of research, given the extensive lead it has, but the U.S. share of the world's scientific research will gradually decline. In fact, considering what other nations are likely to spend in the future, the U.S. share will decline in any plausible scenario of public and private funding of science.

How much of a problem will this be, given that world leadership is an explicit goal of U.S. science policy? What exactly is scientific leadership, beyond the statistical measures of spending and patents? Does leadership provide tangible benefits worthy of taxpayer support? And what about industrial "competitiveness," another frequent policy consideration often linked to scientific leadership? Can or should anything be done about our diminishing standing in world science? If not, how can science policy accommodate the inevitable?

U.S. Leadership, Past and Present

How can we measure world leadership? Ideally, we would like some measure of outputs of research or of scientific competence in different fields, but such information exists only in incompatible bits and pieces.

Total Spending. Leadership, let us assume, is roughly measured by the U.S. share of total spending on research and development. Total spending, of course, is a measure of the input into research, not the output of research. Spending on science is not the same as scientific accomplishment. Questions arise concerning whether a research dollar buys "more" in this country than it does somewhere else. But spending data are reasonably well correlated with the scant indicators of output that we do have, such as publications and citation counts. Moreover, budgets and spending are fairly good leading indicators: if this year's budget finances a big new piece of research equipment, future research activity will probably follow.

Global research and development are so dominated by the industrial nations that it is difficult to assemble data beyond those pertaining to the twenty-six member nations of the Paris-based Organization for Economic Cooperation and Development (OECD). In 1993, the United States accounted for 43 percent of the industrial world's R&D, and this share was more than that of the next four largest performers—Japan, Germany, France, and the United Kingdom—combined. Various data problems involving price comparisons, currency conversions, and consistency of reporting definitions complicate the picture, but not enough to negate this striking picture of world leadership. Without question, any unbiased poll of experts asking which country has the most high-quality scientists, or which country "does the best science," would undoubtedly put the United States in first place. Since the U.S. share of the OECD's gross domestic product is more than a third, the commanding position of U.S. research is not surprising.[1]

Publications, Patents, and Education. Other measures confirm the U.S. leadership position. Since spending on research measures only the inputs to the scientific process, we also need to look at measures of output, scant though they are. In 1993, U.S. researchers published 141,000 articles in the world's influential science and technology journals, 34 percent of the world's article output.[2] U.S. science and technology articles are cited by other scientists well in excess of the U.S. share of the world's publications, consistent with the idea that the United States leads in quality as well as in quantity of publications.³ Data on collaboration in writing scientific articles show that coauthorship centers to a remarkable degree on the United States.⁴ Over the longer term, since Nobel prizes were established, U. S. citizens⁵ have won 44 percent of the prizes in physics, 32 percent of the prizes in chemistry, 49 percent of the prizes in medicine, and 67 percent of the prizes in economics.

Another measure of the results of scientific endeavor is patented inventions, although it is subject to several technical limitations. Most important, since patent systems vary considerably from country to country, it is not considered legitimate to add up patents for the world and calculate country shares. One revealing statistic is that U.S. inventors are consistently granted more European patents than the citizens of any single European country are.⁶

Finally, the United States is the unquestioned world leader in the education of scientists. Universities in this country grant a third of the world's doctoral degrees in the natural sciences and engineering,⁷ which is probably an underestimate of U.S. leadership, once international differences in quality of graduate education are considered. Of the students who leave their own countries to pursue graduate studies in science and engineering, more go to the United States than to any other country.

A well-known caveat is relevant. The United States devotes a considerably larger share of its research funds to military R&D than does any other large country. As a result, U.S. research is, in aggregate, less commercially oriented, resulting in less economic impact per research dollar. There is some commercial spillover, of course, certainly enough to fuel occasional complaints by our trading partners about unfair subsidies by the Pentagon related to aeronautics and other dual-use technologies. But much of the military R&D is very specific to particular weapons systems and falls so far at the development and testing end of the spectrum that some analysts recommend that it be omitted from the national totals.⁸

U.S. Leadership, Future

Will the United States continue as a world leader in science? First, let us take a historical perspective. Just a few years ago, U.S. leadership in science funding was even more pronounced than it is now. In the early postwar years, U.S. dominance was virtually total. In 1961, the first year for which reasonably complete data are available, the U. S. share of OECD research and development was around 71 percent.[9] This share has declined steadily to the current one of around 40 percent, or to a few percentage points less than that if we include estimates of spending by countries not in the OECD.

For decades, this downward trend in the U.S. share of science funding was largely a natural convergence reflecting the gradual recovery of the nations that were wrecked in the war—in other words, Japan and most of Europe. If we consider only the United States and Europe, it would appear that the convergence is now just about complete, with countries spending roughly similar shares of their national product on R&D. The stagnant research budgets of most European countries suggest that a "steady state" has been reached, at least insofar as the traditional leading countries are concerned.

Russia. Russia, as usual, is a puzzle. Its once formidable science enterprise has collapsed since the breakup of the Soviet Union. Scientists, once accorded a privileged status, reportedly go months with little or no pay, and Russian science has become a recipient of foreign aid. Scientific "output" has probably fallen by more than half since the USSR fell apart. While there are some indications of incipient recovery, it will be many years before Russia regains its earlier prominence.

Asia. The future of Asian science looks quite different. By every key measure, Asia appears to be poised for rapid increases in its competence in science and in the volume of scientific research it performs. Clearly, the key prerequisites are in place—economic strength, political will, and human capital.

Asia's economic output is now roughly a third or a fourth of the world's total. Moreover, Asia's GDP is growing much faster than that of the rest of the world, including the member countries of the OECD. Since 1981, according to data from the International Monetary Fund, Asia (excluding Japan) has grown by an average of 7.2 percent annually, compared with just 2.6 percent in the industrial countries. If economic growth were to continue at this rate, Asia's economy would double in ten years, by which time the industrial economies would have grown by less than 30 percent.

While Asia's future growth may be slower than in the recent

past, and many other imperatives will compete for governmental funds, the healthy economic outlook certainly suggests that Asian countries will be able to afford their ambitious plans to improve their scientific status. Surely, it would be feasible for the Asian nations to devote between 2 and 3 percent of their GDP to research and development and let that amount grow at the same rate as their economies. Japan and South Korea are both having economic problems, but they apparently regard R&D spending not as a drag on their budgets but as a way to reinvigorate their economies.

Economic growth brings changes in economic structure. Economies go through stages of growth over the long term, progressively moving up the ladder from agrarian, to manufacturing, to service industries, and to knowledge-based industries. This progression up the hierarchy of industries requires steady increases in the technical competence of the work force and the level of technology used in the products and their manufacture. Asian governments apparently intend to spend the money required to climb this ladder, not wait for change to happen on its own.

The rapid economic growth in East Asia has received considerable attention from economists. Alwyn Young[10] argues that, contrary to widespread belief, rapid Asian growth owes little to improvements in efficiency but is mainly the result of the accumulation of capital, with efficiency still lagging behind that of the West. Since capital is subject to diminishing returns, the capital-accumulation path to growth, Young argues, would become less and less effective. Prospects for improved efficiency suggest a likely Asian economic strategy based increasingly on science and technology, as opposed to sheer capital accumulation.

Japan. Japan's new Science and Technology Basic Plan, released in July 1996, sets out an ambitious program of budget increases and policies to open up the science establishment to operate more efficiently. It deplores the decrease in R&D spending that occurred during the economic downturn of 1993–1994, just the opposite of what is needed to counteract the aging of the population, intensified international competition, and the "hollowing out" of industry. It sets an agenda for improvement of Japan's system of graduate education in science, which is tradition bound and inflexible. The apparent goal of this initiative is eventually to match or surpass U.S. science, both in budgetary and in qualitative terms.

South Korea. South Korea's plan for improvement in science is as ambitious as Japan's. Only thirty years ago, the country's top exports were minerals, rice, raw silk, and cuttlefish. Now, electronic goods are the number-one export, followed by cars and other medium-tech manufacturing products. The next step is obvious: more and better science and technology to support the move to even higher-tech industry. According to the government's latest five-year plan, promotion of science and technology will be looked to as a major stimulus to growth. The plan calls for parity with the advanced countries by early in the twenty-first century.

Based on its R&D spending, Korea is well on its way to achieving this goal. In 1993, the latest year for which comparable figures are available, South Korea's R&D spending at 2.3 percent of its economy put it in the same league as Germany (2.5 percent), Japan (2.9 percent), the United Kingdom (2.2 percent), and the United States (2.7 percent). By now, South Korea has probably passed a few more European countries in this measure of national R&D effort. Like Japan's, Korea's plans appear to be strongly influenced by the success of America's research universities, which both countries would very much like to emulate.

China, India, and the NICs. The situations in China and India are in many ways unique, but both are consistent with the prospects for strong growth in spending on scientific research. China is so populous that it will become a major economic power even if it achieves only a fraction of Western productivity levels. While most data on the Chinese economy are untrustworthy, there is no doubt that the country's economy has grown rapidly in recent years. If China could keep growing at around 7 percent (which would be a mild slowdown), its economy would approach the size of ours in twenty years or so. Of course, the Chinese economy would have a much different structure, but it would have the capacity to play in the major leagues of science funding.

India is likewise a country so huge that even low per capita production generates a large national economic output. India has long put substantial resources into higher education, possibly more than might be clearly justified by economic needs. The result is a large and rapidly growing supply of scientific and technical workers.

The other "newly industrializing countries" of Asia—Hong Kong, Singapore, and Taiwan—are also stepping up their science spending. At present, their science budgets total just a few billion dollars per year, so even with rapid growth they will in no way challenge the United States. Still, they will be able to marshal a sizable array of scientific talent and facilities, and quite possibly they will, among themselves, achieve near-leadership positions in one or more subfields of science.

Higher education. Data on the awarding of doctorates provide another perspective on the growth of scientific capabilities in Asia. According to the National Science Foundation, in 1992 Asian universities produced 20 percent of the world's Ph.D.s in science and engineering, and this share is steadily increasing.[11] More than 5,000 Asian students earned doctoral degrees in science and engineering in the United States in 1993. Data are sketchy on how many of these graduates return to their home countries, maybe half.[12] Asian universities' production of bachelor's and master's degrees in science and engineering fields is also very large and growing rapidly.[13]

Looking Ahead a Few Years

Let's take another look at the U.S. share of world science spending, adding in some other countries and looking to the future. To generate a five-year projection,

- We include in our calculations the science spending of China, India, and the rest of Southeast Asia.
- We assume that U.S. R&D spending stays the same in real terms, with the idea that private sector increases will offset decreases in the public sector.
- We assume that the rest of the OECD's science spending, led by that of Japan and Korea, grows by just 3 percent annually.
- We assume that the other countries in Asia increase their spending by 5 percent annually, which is somewhat less than what their plans call for.

According to these assumptions, the U.S. share of world spending in 1993 was not 43 percent but, including the non-OECD countries' science spending, around 36 percent.

By 2002 the U.S. share would be around 26 percent of the total (figure 10-1).

This trend may continue for some time. As we look ahead fifteen years from now, according to recent analysis by RAND

FIGURE 10–1
U.S. Share of World R&D Expenditures, 1961, 1993, and 2003

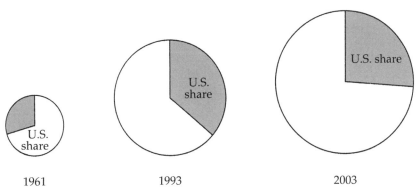

NOTE: Area of each circle represents volume of world R&D.
SOURCE: OECD and author's estimates.

Corporation,[14] the gross domestic product of China will be more than double Japan's and equal to that of the United States. The total GDP of China, Japan, India, Korea, and Indonesia will be double that of the United States. This prediction suggests a continuing buildup of research capacity in Asia, with the U.S. share steadily sinking, unless research spending resurges in the United States.

With the U.S. share of science spending falling well below 30 percent, it seems unlikely that leadership will continue, at least not in the sense of having more and better science in every field. The United States will, for many years to come, remain superior to any other country in science in general. It might even continue to be the leader in each of the main fields of science: chemistry, physics, and so forth. But surely other countries will move ahead in particular subfields of science or will build some facility that is better than any other in the world. And maybe the relevant comparison is not the United States with some particular other country but U.S. science compared with the aggregate in the rest of the world. In that case, leadership will certainly decline.

Implications of Relative Decline

Scenarios of relative decline are nothing new. They go back to Spengler and earlier. Most recently, in the late 1980s, new pronouncements of U.S. decline emerged, epitomized by the best-selling 1987 book *The Rise and Fall of Great Powers*, by Paul Kennedy.[15] He and other "declinists" argued that the era of U.S. dominance in world affairs was coming to an end, with grave implications for American power and influence. The argument was echoed in any arena for which statistics were available: manufacturing output, exports, wealth, and income. Just as we have done in this chapter, declinists would typically present a data series starting at the close of World War II, when the United States enjoyed the lion's share of any particular good, and would trace the hyperbolic downward path of that series to the much inferior present, where this nation is merely first among equals.[16]

The problem does not lie so much with the data as with their interpretation and the ensuing policy recommendations. Essentially, the declinists say that the United States should give up its futile and self-defeating efforts to exercise global power and accept the inevitable by drawing back and adjusting gracefully to its dotage. But the diagnosed problem (future collapse) was disproportionate to the cause (military spending). The declinists believed that the 4 or 5 percent of GDP devoted to military spending in the mid-1980s was unsustainable and would lead ultimately to America's ruin. But that amount was clearly affordable, and in any case has fallen considerably since then, to about 3 percent in 1997. As with military leadership, so too with international leadership in science: if the United States really wants leadership, it possesses the economic capacity to achieve it. Maintaining federal science spending at, say, a little more than 1 percent of GDP, and maintaining a good environment for industrial R&D, are all it would take.

Science Policy and the Price of Leadership. The discussion of the relationship between world leadership in science and funding scientific research is important. If the United States chooses to relinquish its relative standing through changes in budgetary priorities reflected in reduced science spending, we need to understand the consequences of that loss. What are the specific costs of a reduced leadership role? How much would it cost to prevent that decline, and would it be worth doing? If we can particularize the benefits of scientific leadership, can we find an alternative way to achieve those benefits? While a diminished share of the world's best science may be inevitable, the United States still has scope for positive policy steps to avoid the harmful consequences of that relative decline, as well as for precluding counterproductive reactions to it.

Trade policy is an interesting parallel. As the U.S. share of world manufacturing output declined in the decades following World War II, particularly in steel, textiles, automobiles, and machine tools, trade policy took on various shades of protectionism, export promotion, and industrial policy. One can imagine a science policy version of protectionism coming into being. While tariffs and quotas have no exact counterpart in science policy, the United States could drift toward a kind of isolationism, whereby federal funding would be reserved strictly for the most inward-looking science projects and U.S. scientists were not supported in their efforts to collaborate with foreign scientists. One could also imagine a negative public reaction to the large numbers of foreign graduate students that study in the United States.

Not all these potential reactions are necessarily bad policy. In fact, foreign companies are becoming more and more able to appropriate the results of basic research performed elsewhere. What, some ask, is the public benefit to supporting technology development or research in cases where many of the spillover benefits help foreign countries? Is avoiding that appropriation by foreigners not the perfectly understandable motivation for corporations to devote fewer resources to basic research and more to applied research and development related expressly to the corporation's own products? Instead, as foreign science improves, should we not spend more money on staying abreast of new developments and applying the latest foreign results to our own science and technology purposes?

Thus, science policy is faced with a dilemma. With more good science around the world, more knowledge will be produced and more use made of it. Clearly, the situation we want to achieve is one of continued incentives to perform basic research with the results disseminated widely. But that desirable situation will come to pass only in a system of cooperation and partnership.

How Much Is World Leadership Worth?

As we have seen, we have no obvious standards for setting science budgets and allocating federal funds among the fields of science. Even the rate-of-return analysis, which may appear to have possibilities for quantitative budget management, offers little help, since it applies to the totality of spending and not to spending on the margin or to newly proposed, discrete projects.

Leadership Defined. World leadership is one of the quasi-objective criteria that have been proposed for allocating funds. A committee of the National Academy of Sciences[17] proposed, for example, that the United States should spend "enough so that we are the world leader in selected fields, and have world-class abilities in all areas." A goal of the National Science Foundation is to "enable the United States to uphold a position of world leadership in all aspects of science, mathematics, and engineering."[18]

That definition of leadership is something like a corporate mission statement that says, "Our company will be the leader in industry X." It is not immediately clear just where such a goal comes from or how it can be justified in dollars and cents. Management experts warn against this sort of one-upsmanship approach, since letting competitors set the agenda may cause a company to ignore more innovative strategies. Science leadership is also hard to measure. While a company could measure leadership by its share of the industry's sales, determination of international leadership in, say, mathematics would have to be based on more subjective criteria such as expert opinion.

Advantages of Leadership. There may well be some special benefits of leading the world in a particular field of science. Perhaps a disproportionate share of broad social benefits accrues to the world leader rather than the runner-up. In other words, a given level of scientific achievement may be more valuable if the country is in first place rather than second. This advantage probably happens with unpatentable new fundamental knowledge that benefits its discoverers only for the limited time before others catch on. We have no quantitative evidence on this question, however.

Another likely advantage of leadership, or at least of near-leadership, is that one's scientists are capable of recognizing and taking advantage of new discoveries elsewhere in the world. An obvious example would be research related to DNA: a huge discovery was made, but the development and application depended on the ability of scientists worldwide to understand what had been discovered and to carry it forward. In another example, high-temperature superconductivity was discovered in IBM's Zurich laboratory but exploited by a group at the University of Houston who were working on superconductivity. Those scientists were able to modify their apparatus and, more important, their perspective, when they read about the foreign discovery.

Scientific advances are usually not the result of spectacular single discoveries like these but of a linked sequence of smaller results that lead to something new. Often those results emerge from different laboratories and involve many scientists who stay in contact. Therefore, it is argued that domestic scientists need to be competent to pick up on relatively small steps made elsewhere, develop the research, and contribute to advances. U.S. scientists must be sufficiently competent to be "fast followers." In addition to building on the discoveries of others, domestic groups must maintain their membership in the international networks of leaders.[19]

Problems with Leadership. The "world leadership" approach, however, can create problems if it is implemented too freely or when strict economic criteria are set aside. The leadership argument was used in the 1970s, for example, to justify federal funding of a supersonic passenger jet. Had it been produced, the plane would have been an enormous money loser, regardless of any technical superiority over the supersonic French-British Concorde or the Soviet Tupolev TU-144, both of which were unprofitable.

Earlier in the century, luxury ocean liners were an object of international competition for technological leadership, determined by how fast they could cross the Atlantic. The lead changed hands several times between 1910 and the 1950s. The final "winner," with a record average speed of 38 knots, was the *United States*. But within just a few years the *United States* and all such transatlantic liners were made technically and economically obsolete by jet airliners, and thus this particular contest for technological leadership simply disappeared.

More recently, the superconducting supercollider, which would have cost more than $11 billion, was billed by its proponents as necessary to maintain the United States as the world's preeminent scientific nation. The project was canceled after several billion dollars were spent; it is too early to tell whether this decision will cost the United States its leadership in any particular field.

The problem with leadership as a guide to resource allocation is that it may conflict with the goal of achieving a high rate of return on a nation's scientific investments. Suppose that of two fields of science, A and B, the prospective rate of return is substantially higher in A because exciting new possibilities have suddenly opened up. But suppose it is judged that the United States is already a leader in A but somewhat behind in B. The "world leader" criterion would direct investment into field B, even if the scientists in that field were already well provided for or even if they did not have any prospects for using the money productively.

Another possible problem occurs when budgets become so tight that across-the-board cuts at the margin are no longer feasible and some particular field has to be eliminated. This predicament is becoming more frequent at individual universities and laboratories and is likely to be a serious if unpalatable option at the federal level. The world leadership criterion, however, would either cease to be relevant or else might call for incremental cuts everywhere to keep each subfield viable.

Thus, leadership is good to have and is clearly worth something, but it may be hard to identify and increasingly costly to achieve in all areas. It is far more relevant in basic science, where the national competency of physics, for example, is being judged, than as a rationale for particular big projects. Indeed, it is dangerous to use leadership as the justification for some particular artifact of big science, like the supersonic transport or the world's biggest particle accelerator.

That Old Shibboleth "Competitiveness"

The press has often portrayed scientific achievement as a battle for international dominance. In 1957 a *Newsweek* cover story read "World War of Science—How We're Mobilizing to Win It." The "growing army of Soviet scientists" was the clear enemy. In the 1980s, Japan had become the new science and technology antagonist, and *Newsweek* now pictured a samurai warrior lunging out of a computer.[20] Later, the press saw Japan's putative lack of proficiency in fundamental research as favorable to U.S. interests.

In recent years the term *competitiveness* has been invoked as an objective of international economic policy and of science policy as well. The diagnosis of lagging competitiveness often leads to a variety of interventionist cures, including thinly veiled protectionism, industrial policies, export promotion, and so forth. Growing R&D

achievements in the very countries perceived as this nation's chief economic competitors, along with tighter science budgets, provide ammunition for those who argue the case for science funding.

Economic competitiveness has a wide variety of meanings. To some, it is a broad term used as shorthand for the general health of industry. The Council for Competitiveness, one of the several organizations whose name incorporates the word, defines it as "the capacity of a nation's goods and services to meet the test of international markets while maintaining or boosting the real incomes of its citizens." In this context, it connotes a sizable export sector and the absence of a trade deficit, goals not to be achieved by means of low wages or downsized work forces but rather by improvements in efficiency and technology.

Others seem to view the world as an economic battlefield. The United States, some writers have asserted, is losing out in the head-to-head battle for economic supremacy with Japan, East Asia, the developing world, and various other nations that are exporting more than they import. Some view economic competition as the replacement of the cold war, the new battle for national survival. Some would even have the CIA spy on economic competitors (including NATO allies) in support of this newly defined competition. As in the more benign interpretation, improvements in industrial technology are part of the winning strategy.

In the view of many economists, however, *competitiveness* is a foggy term that is unhelpful in designing efficient policies and indeed is likely to foster bad policy. This is not the place to repeat the mainstream economic criticisms of competitiveness lore, best articulated in the writings of Paul Krugman,[21] but only to encapsulate their message as it applies to science policy. The idea that international trade is a battle, or is in any way an economic version of the cold war, is economic nonsense. In broad terms, international trade is mutually beneficial to nations that trade with each other, and indeed the postwar economic boom has been made possible by the opening up of trade channels that had been closed in the 1930s by tariffs and in the early 1940s by war.[22]

Implications for Policy

The essential change is this: in the future there will be a two-way flow of information between the United States and the rest of the world, in contrast with the days when the United States was doing more than half the scientific research performed worldwide and most of the information was flowing out from the United States. Whatever the implications for U.S. leadership per se, these developments have favorable aspects. Is it not a good thing for more scientific discoveries to be made, particularly if international spillovers bring benefits to the United States? Maybe the United States can again be a "free rider" on research done elsewhere, as it was in preleadership days.

But such spillovers, or transmission of knowledge, do not happen automatically. Making sure that knowledge flows freely should be the basis of our policy reaction to these developments. In a world where more good science is being done in many locations around the world, communication is vital to any individual scientist's own work. Most U.S. researchers are well aware that they have to keep abreast of leading research, and indeed the data on international coauthorships indicate that cross-border collaboration is steadily rising. (That development is also an indication of the ever-increasing competence of foreign science.) Technical advances such as the Internet are making such communication easier. If Thomas Edison were trying to invent the incandescent light bulb today, he would want to scan the World Wide Web to see who else was making progress: "Let's see—*Infoseek, search for*, filaments, carbonized bamboo. Hmm, no hits. I guess we'd better try that in the lab." Channeling more federal money into encouraging such international contacts would be worthwhile.

Industrial firms are doing less basic research than they used to, in part because they find it harder to exploit it quickly with the growing ability of foreign researchers to make use of U.S. results. U.S. firms will find it increasingly profitable to do likewise and troll the rest of the world for research results, whether by establishing research units abroad or paying others to look for research. This approach may become a new consulting specialty, the service of reporting on world developments in very specific areas. (Such "reporting" may, of course, shade into "industrial espionage.")

Government-to-government relationships will be more problematic. We can divide the issues according to whether the knowledge gained by the research becomes intellectual property or enters into the world's open source knowledge bank. U.S. trade negotiators over the years have done a good job of focusing on intellectual property issues as part of trade in goods, where before the question had usually been whether patents or copyrights were being infringed on. As industrial research grows worldwide, there will be more such property to protect and more of a two-way street. We may well see complaints about U.S. infringements of *foreign* patents. In a world with more evenly distributed intellectual property, other nations will have a greater stake in a system of rules.

What will happen to "big science," the fundamental research conducted in major facilities financed primarily by the United States? We have progressed from the stage of "going it alone" (early space and nuclear research) to more recent attempts at international cost sharing, where we now contribute money to a limited number of foreign projects so that our scientists can participate in using those facilities. In the future, it will become more common for the United States to collaborate in foreign big science projects.

A prototype controversy involves U.S. participation in the construction and use of the planned Large Hadron Collider (LHC), a $6 billion facility to be built in Geneva by a consortium of European nations with assistance from "nonmember" countries such as the United States and Japan. The Energy Department tentatively agreed to contribute $450 million to the LHC for construction of the circular tunnel and work on detectors, but that plan met with congressional criticism that the DOE was being too generous. Supporters of the plan pointed out that the United States would pay about 10 percent of the project cost but that 20 percent of the scientists who would work there would be American. Finally, a mutually satisfactory agreement was reached, but the contentious issues that arose there will no doubt surface again with other such projects.

Beyond this sort of arrangement, we need to anticipate the day when foreign nations strike out on their own and exclude U.S. scientists from the use of new facilities, putting them at a disadvantage in using the results. To date, other nations—Japan included—seem reasonably willing to let U.S. scientists have access to their new research facilities so long as the United States pays its "share." Yet we need a structure that will facilitate such agreements and better define in advance what a nation's share of the cost should be. How much if any of the fixed costs should foreign nations contribute, for example, and how much should they give toward the operational costs directly related to their scientists' participation? One model is the way the European Community plans and uses its research facilities. Another model, only partially relevant, is the arrangement whereby U.S. universities share and manage certain national laboratories. Whatever the system, one thing is certain: international cooperation must extend beyond the OECD to include other countries, particularly those Asian countries that are making a big push in science.

Conclusion

In light of budget realities and the growth of science elsewhere, world leadership in every field of science may no longer be a realistic goal. To a great extent, the erosion of U.S. leadership was inevitable, the result of long-term trends in other nations and particularly the growth of science in Asia. The United States could, without undue strain on the federal budget, come closer to world leadership than the projections here indicate, but that seems unlikely. The picture is not uniformly bleak, though, with the global expansion of science bringing rapid growth in the world's knowledge base. We should welcome this explosion of new knowledge as a benefit to the United States, not fear it as a threat.

11
The Price of Downsizing Science

SUPPOSE THAT EVERYTHING goes according to plan. Suppose that the federal science budget is reduced more or less in accord with the 1997 projections of the American Association for the Advancement of Science, so that by 2002 nondefense funding for research and development is down 9.4 percent in real terms from 1997 and 14 percent from 1994. Suppose defense R&D drops by even more. Even suppose that the science budget stabilizes after 2002 and climbs slowly back toward the spending levels reached in the early 1990s. We do not know with any certainty that this is the scenario that will play out, but it is what our elected officials say they want to do. What consequences would we notice, say, ten years from now?

Examining macroeconomic statistics at that future time, we would probably not see much effect. It would be too early to see a measurable slowdown in productivity growth, although one might be starting to develop. Looking at progress in scientific knowledge, the situation might also appear satisfactory. Progress in biotechnology and information technologies will have advanced considerably, because they will still be priorities for the federal and, more important, the industrial budgets. There might well be huge breakthroughs in preventing and treating cancer, AIDS, and some other diseases. Any number of new technologies will have been launched in the form of new goods and new production techniques. Maybe U.S. industry will contrive to do enough "free riding" (taking advantage of foreign advances in science) to ease the productivity lag somewhat.

Something will be missing, though, some undiscovered new knowledge, but of course we will have no idea what that knowledge might have been or how many useful applications it might have had. For example, worldwide destruction of natural habitats and species will probably accelerate with the relentless expansion of farmland, but few may connect this result with the lack of funds for agricultural research that would have made existing farms more productive. Maybe research on prostate cancer will still lack resources needed to advance substantially. Similarly, we can only speculate on other diseases not cured and scientific breakthroughs not made.

Although most of today's federal laboratories will still be in operation, many will be substantially smaller, and productivity and morale in their aging work force will have reached new lows. Congressional committees will hear testimony that oversight needs to be simultaneously strengthened and streamlined and that the laboratories' missions need to be clarified. While not much science will be coming out of the labs, no doubt they will have maintained their capacity to produce nuclear weapons, and they will possibly have made progress in cleaning up nuclear wastes. Their connections with industry will be minor. NASA's miniaturized budget will support miniature space probes but not manned space flight beyond earth orbit.

U.S. industry will be doing reasonably well, approximately in accord with long-term trends. More research will be farmed out to corporate or university laboratories overseas. A new field of industrial consulting will develop: scientists will cover foreign developments and report profitable information back to companies and industry consortia. Miscellaneous technology-subsidy programs will still exist, but they will be too small to have any measurable impact on industrial research generally, let alone be cost effective. Various regulations mandating technology change will be touted as the means to make U.S. industry more "competitive." One possible technology policy issue will be whether slow-selling electric cars should be subsidized even more or whether lower prices should be federally mandated.

Science policy makers will no longer proclaim world leadership to be the nation's goal. The operative concepts may instead be "achieving and maintaining parity in key science fields" or "strengthening the global science alliance" or "learning from the leaders." Perhaps, for instance, some Asian nation will design an important new research facility for its exclusive use, denying U.S. scientists access, demanding unreasonable cost-sharing arrangements, or absorbing results of U.S. research in an overly aggressive way (headline: "Administration Accuses China of Strip-mining U.S. Science").

Universities will be noticeably different after ten years of diminished federal funding. Today's leading research universities will still be excellent, but they may have shrunk a little. Science faculties will have aged, owing to universities' diminished ability to hire young scientists. A number of science and mathematics graduate programs and some departments at the less prominent universities will have been eliminated. Production of science Ph.D.s will be lower, with both fewer foreigners and fewer U.S. citizens among the degree recipients. Greater numbers of U.S. graduate students and postdoctorals will need to go abroad to find expertise in certain specialized fields.

Some universities will replace the absent federal funds with general funds, favoring those scientists who have the prominence and influence to acquire such support. Corporate funds will fill only a small portion of the gap left by downsized federal funding. Corporate money, of course, will flow to the most industrially relevant fields, with emphasis on applied rather than basic research. In certain fields, mainly biotechnology and information technology, faculty members will increasingly form their own research corporations, making their money by selling rights to the intellectual property they discover and stake out. Applied (that is, marketable) research will often crowd out fundamental research. Tuition for undergraduates will have been raised still again, possibly in response to new federal financial aid to students, and this income will provide some small cushion for research budgets.

Altogether, this is a picture of diminished excellence, with increasingly evident traces of mediocrity. Words like *disaster* or *calamity* would be overstatements, except to those with strong convictions that the golden age of U.S. science should have lasted longer. Moreover, with the depleted science infrastructure that exists in the downsized future, it will be difficult to catch up with what the nation could have achieved under an alternative scenario of healthy scientific growth, 1997 to 2007, should future policy makers wish to make R&D a priority once again.

And, yes, the federal deficit will still be with us.

What Are the Alternatives?

Maintaining U.S. excellence in science is not an impossible dream, like ridding the world of crime or narcotics. It is not a problem of exploding costs, as with social security or Medicaid. It does not even require building vast new infrastructure or creating new institutional arrangements; most of what we already have is sufficient. The gap between excellence and mediocrity is measurable in the low billions of dollars per year. Clearly, continued excellence in science is within the nation's reach.

Specifically, what is required? Our superb research and innovation system has not yet been severely weakened. It seems indisputable that the United States can afford to take today's science budget as a baseline, comb out some of the questionable subsidy programs, and increase spending by around 3.0 or 3.5 percent annually (in real terms), equal to or just a bit faster than the long-term growth of the U.S. economy. I am not proposing an off-budget trust fund for science or any kind of "entitlement" for scientists but simply suggesting the dimensions of federal funding needed to achieve reasonable goals in science and technology. This formula bypasses the never-settled question of whether the current budget is the right size but does presume that R&D should at least keep up with economic growth, not fall behind.

I do not mean to overstate the case for science. Taxpayers do not "owe" exponential growth to the scientific establishment. We should not raise the R&D budget at some unsustainable rate that would set us up for still another fall later on. The basic idea should be to increase science support in a predictable fashion, slightly faster than economic growth, and to reassess the situation every five years or so. That R&D would grow as a portion of gross domestic product is in accord with historical trends whereby growing economies steadily become more technologically oriented.

But in any case, this approach has not been accepted; there are too many other appealing ways for Congress to spend money. Indeed, the August 1997 budget package, for all of its bulk and political fanfare, was little more than a collection of new transfer payments and tax cuts that contained virtually no elements of supply-side stimulus or long-term social investment. As was the case with the budget proposals discussed in chapter 2, R&D funding played no discernible role in the debate. Therefore, let us consider how to cope with the more likely future characterized by limited resources for R&D.

Target federal funds more effectively. Several times we have been led to the rather obvious conclusion that science spending could be cut without great harm if we could defund the least productive programs. But how should we go about that?

The status quo is a vast array of science funding programs, each of which has latched onto the increasingly automatic yet logically invalid justification that "some science gives large benefits, program X is science, therefore program X gives large benefits." This approach has engendered skepticism and indiscriminate funding on Capitol Hill. Moreover, it has reduced the overall rate of return to be gotten from the science budget.

The problem lies not so much with the total of science spending but with how it is allocated. The United States spends far more proportionally on health, defense, and the environment than any other nation. This may be a good decision, but it is not really a "decision" at all. It is rather the outcome of a mysterious, convoluted political process with many diverse, sometimes conflicting, goals. Is a disorganized political process likely to produce better results than a conscious, institutionalized budget process for science?

One reasonable attempt to organize science budgeting better was presented in a 1995 National Academy of Sciences report, *Allocating Federal Funds for Science and Technology*.[1] While the report is overly optimistic in its hopes to involve the president in detailed decision making, it does outline a reasonable approach to getting a handle on the diverse and inadequately coordinated panoply of R&D programs. It recommends the development and use of what it calls a federal science and technology (FS&T) budget, the essential feature of which is the establishment of a unified science budget.

In addition, the report defines FS&T to include only the R&D spending that is devoted to expanding fundamental knowledge and creating new knowledge. This classification would exclude funding of production engineering, testing, and upgrading of weapons systems, most of which is in the Defense Department budget. In other words, it would exclude close to half of what we usually think of as federal R&D. Some have declared this to be a bad (that is, federal money-losing) strategy for science, arguing that reducing the apparent size of federal science would degrade science funding in the eyes of Congress. The authors of the report argue that the FS&T budget is more realistic, excluding as it does the sort of federal funding that most would agree is not really what we should mean by research and development.

Directing funds to the best science requires information that may be costly or difficult to produce. For example, the Government Performance and Results Act (GPRA), the 1993 law by which agencies are required to evaluate the "outcomes" of their programs, has the potential of generating excessive paperwork at substantial cost. While the objective of GPRA—to determine whether programs are working—is unquestionably worthwhile, it will be difficult to achieve. For one thing, determining the contribution of specific funding to the world of science is a huge chore. Moreover, the evaluations that agencies perform on themselves may be assumed to be biased in favor of their own programs and will therefore lose credibility. Recalling that one of the criteria for federal funding was the ability of an agency to deliver funds efficiently, we need to add the cost of doing GPRA evaluations into the overhead costs of each agency. To date, no GPRA analyses have been satisfactorily completed, so it is too early to judge how well this act will improve decision making.

Finally, we need to recognize the importance of stability in science funding. The 1995 NAS report called for two-year budgets, which would be one good step in the direction of stability, or at least an aid to planning. (This reform has been suggested many times over the years, with zero results.) Science is a long-term investment, and many aspects of it demand long-term planning. Students, for example, would like to know that there will be a market for their services before undertaking the long, arduous path to the Ph.D. It is not that students need special treatment but that uncertainty will steer potentially good scientists into other careers. As noted in chapter 9, the tax credit for R&D has lost effectiveness owing to the uncertainty over its continued existence. Greater stability in funding can make up for some of the downsizing.

Nurture—but do not directly subsidize—industrial R&D. With federal downsizing, we should encourage the private sector to take up the slack wherever possible. Industry, of course, will not duplicate federal programs one for one. What is achievable, however, is a healthy response from private funding that will accomplish many of the same goals. If, for example, the Advanced Technology Program were to be eliminated, industry might pick up a number of the projects, particularly given the strong presumption that much of the program displaced private funding. The savings could help fund research with broad social benefits that would otherwise be lost.

Industrial research is growing, but we still need to surround it with policies that will provide a favorable climate for research and innovation. If we see industrial R&D slow down, we need to look not to subsidies but to the more subtle and indirect causes, such as tax policy as it affects the effective rate of return to investment in R&D. Also among the usual suspects are federal regulations of all kinds, which are becoming ever more complex and, in their labyrinthine ways, more burdensome. The whole range of regulations, health care policies, and complex legalisms that surround pharmaceuticals, for example, probably do more to affect the rate of return on R&D than ten Advanced Technology Programs would.

The final three recommendations can be stated more briefly:

- *Rationalize the roles of the federal laboratories.* Assuming a downward trend in funding, it is especially important to define the goals of the federal laboratories, eliminate wasteful and duplicative programs, and improve efficiency. First on the agenda should be a high-level effort to define the goals, but not a commission to target labs for elimination. The survival of a highly valuable but misused resource is at stake.
- *Recognize the growth of science worldwide and develop policies to take advantage of it.* In light of budget realities and trends in science spending abroad, leadership does not make a lot of sense as a goal of science policy. It is not realistic in the aggregate, it is not persuasive, and it does not help in establishing priorities. With a downsized science enterprise, the United States will need to be increasingly interactive with the countries that have chosen to push ahead in science more rapidly than this country.

- *Protect the federal support for academic research.* This book has argued that research performed by universities best satisfies the criteria for government support. It delivers broad social benefits, it is unlikely to be financed sufficiently by other means, and the federal mechanism for allocating the money is reasonably efficient. Most of the costs of downsizing science would ensue from the decline of academic research.

Reflecting on the Future

This book builds on two particular strands of economic research: (1) measurement of the economic effects of R&D, as developed in the works of Solow, Mansfield, and others; and (2) the theory of public goods and intellectual property. In my view, the first type of research has gone about as far as it can with the existing data, which are virtually mined out.[2] The theory of intellectual property, however, is still evolving. In this study we have encountered several areas where the theory needs to be developed and applied. How will academic researchers respond, for example, to opportunities for patenting under alternative legal frameworks? How should cooperative research agreements be structured to foster productive partnerships among federal labs, universities, and industry? How should international research cooperation be structured? More generally, what institutional framework for property rights will generate the incentives that maximize the benefits from scientific research during a time when public resources are limited?

Notes

CHAPTER 2: END OF THE GOLDEN AGE

1. Thomas Malthus, *An Essay on the Principles of Population*, 1798. Malthus said that there was a tendency for population to grow faster than food production and that starvation at the margin would control population growth.

2. Vannevar Bush, *Science—The Endless Frontier: A Report to the President on a Program for Postwar Scientific Research*, July 1945 (Washington, D.C.: National Science Foundation, 1990), p. 5.

3. There is no "science budget" as such. The projections presented here come from the American Association for the Advancement of Science (AAAS), which converts proposed agency budgets into projected R&D authorizations. AAAS issues new projections with each new budget proposal by the administration or Congress.

4. National Science Foundation, *National Patterns of R&D Resources: 1996* (Washington, D.C.: National Science Foundation, 1996), table C-3.

5. For example, reductions in the estimates of future inflation automatically increased the estimates of future spending in real terms. The actual budget legislation—the Taxpayer Relief and Balanced Budget Act of 1997—made use of even more favorable economic assumptions and projected a budget surplus by 2002. Required reductions in discretionary spending, however, were left largely unspecified. For this reason the AAAS was unable to look at the implications of this budget agreement for the R&D funding agencies.

6. See Congressional Budget Office, *Long-Term Budgetary Pressures and Policy Options*, March 1997. See also Congressional Budget Office, *The Economic and Budget Outlook: An Update* (Washington, D.C.: GPO, 1997).

7. James Kitfield, "Biting the Bullet," *Government Executive*, vol. 29, no. 1 (January 1997), pp. 18–22.

8. Albert H. Teich, Testimony before the House Committee on Science, July 23, 1996.

9. Adjusted for inflation, which was assumed to be 2.6 percent during the coming year.

10. *AAAS Report 22: Research and Development FY 1998*, March 24, 1997 (hereafter AAAS report). These totals incorporate an adjustment by AAAS, a reduction of $883 million that appears in the FY 1998 budget for DOE but that is reserved for project costs in FY 1999 and later.

11. This argument was heard from some congressional sources in defense of the balanced-budget plan. This is also the thesis of Terence Kealy's book, *The New Economics of Science* (New York: St. Martin's Press, 1996), about which more later.

12. Donald Stokes, *Pasteur's Quadrant: Basic Science and Technological Innovation* (Washington, D.C.: Brookings Institution, 1997).

13. National Academy of Sciences, Committee on Criteria for Federal Support of Research and Development, *Allocating Federal Funds for Science and Technology* (Washington, D.C.: National Academy Press, 1995).

14. At one point a science agency's budget submission used the unfortunate term *curiosity-driven research*, a term that has since become taboo in describing federally supported research.

15. It has been said that of all the scientists who have ever lived, 90 percent are alive today.

16. Julian M. Alston and Philip G. Pardey, *Making Science Pay: The Economics of Agricultural R&D Policy* (Washington, D.C.: AEI Press, 1996), pp. 11–13.

17. The classic article in this field was Zvi Griliches, "Hybrid Corn: An Exploration in the Economics of Technological Change," *Econometrica*, vol. 25 (October 1957), pp. 501–22.

18. Estimate made by J. D. Bernal, cited in Harvey Brooks, "Evolution of U.S. Science Policy," in Bruce L.R. Smith and Claude E. Barfield, eds., *Technology, R&D, and the Economy* (Washington, D.C.: Brookings Institution and AEI Press, 1996), p. 15.

19. National Resources Committee, *Research: A National Resource*, vol. 1 (Washington, D.C.: GPO, 1938), p. 178.

20. National Science Foundation, *National Patterns*, table C-3.

21. AAAS report.

22. Beginning in 1992, the sample of companies in the survey of R&D was expanded to include more nonmanufacturing firms and more small firms. Data for 1988–1991 were adjusted to reflect the sizable upward shift in the series that apparently resulted from the changed sampling. It would have been desirable to adjust pre-1988 data, but there was no obvious way to extend the adjustment that far back in time.

23. The source for these numbers is given in chapter 10. It should be kept in mind that even a downsized U.S. science enterprise would still be the largest of any single nation.

24. David Goodstein, "The Big Crunch," NCAR 48 Symposium, Portland, Ore., September 19, 1994.

25. See National Science Board, *Science and Engineering Indicators, 1996* (Washington, D.C.: National Science Foundation, 1996), chap. 3, for an analysis of the labor market for scientists and engineers, based on data from federal surveys and from professional associations of scientists, such as the American Mathematical Society and the American Institute of Physics.

26. See Linda R. Cohen and Roger G. Noll, *Research and Development after the Cold War*, Publication no. 431, Center for Economic Policy Research, Stanford University, July 1995.

27. Dwight D. Eisenhower, *Public Papers of the Presidents*, "Farewell Radio and Television Address to the American People" (January 17, 1961).

CHAPTER 3: MEASURING THE COSTS

1. For a good review of the empirical results and the associated issues, see National Science Board, *Science and Engineering Indicators, 1996* (Washington, D.C.: National Science Foundation, 1996), chap. 8. Also see Richard R. Nelson, *The Sources of Economic Growth* (Cambridge: Harvard University Press, 1996), chap. 1.

2. Nowadays, most writers dismiss this "linear model" as outmoded and sometimes present a complex array of interconnected boxes and arrows depicting all the feedbacks and complications in the way science "really works." It seems reasonable, however, to start with this simplified approach, recognizing that it does not fully describe the complex process of innovation.

3. Robert M. Solow, "Technical Change and the Aggregate Production Function," *Review of Economics and Statistics*, vol. 39 (1957), pp. 312–20.

4. Edward F. Denison, *The Sources of Growth in the United States and the Alternatives before Us*, Supplementary paper no. 13 (New York: Committee for Economic Development, 1962).

5. Edward F. Denison, *Why Growth Rates Differ: Postwar Experience in Nine Western Countries* (Washington, D.C.: Brookings Institution, 1967), p. 298. Another pioneer in this field of research was John Kendrick; see, for example, his *Postwar Productivity Trends in the United States, 1948–1969* (New York: Columbia University Press, 1973).

6. U.S. Department of Labor, Bureau of Labor Statistics, Press release 95-518, January 17, 1996.

7. Some of the key studies are cited later in this chapter. For a recent survey of the field, see Bronwyn Hall, "The Private and Social Returns to Research and Development," in Bruce L. R. Smith and Claude E. Barfield, *Technology, R&D, and the Economy* (Washington, D.C.: Brookings Institution and AEI Press, 1995).

8. Changes on the margin do not have a clear interpretation either, as we shall see later in this chapter.

9. The real price of computers was estimated to be only 4 percent as high in 1988 as it was in 1973. Computers are a producer good and therefore figure in the estimate of the capital stock. If investment in computers is underestimated, then Solow-type measures would give an overestimate of the contribution of technical change.

10. *Toward a More Accurate Measure of the Cost of Living*, Report to the Senate Finance Committee by the Advisory Commission to Study the Consumer Price Index (typescript), December 4, 1996.

11. Leonard I. Nakamura, "Is U.S. Economic Performance Really That Bad?" Working paper 95-21/R, Federal Reserve Bank of Philadelphia, April 1996.

12. Zvi Griliches, "Research Costs and Social Returns: Hybrid Corn and Related Innovations," *Journal of Political Economy* (1958), pp. 419–31.

13. Edwin Mansfield, "Contributions of New Technology to the Economy," in Smith and Barfield, *Technology, R&D, and the Economy*, pp. 114–39.

14. M. I. Nadiri, "Innovations and Technological Spillovers," Working paper no. 4423 (Cambridge, Mass.: National Bureau of Economic Research, 1993).

15. *Survey of Current Business* (U.S. Department of Commerce), November 1994, pp. 37–71.

16. As discussed in chap. 5, the "excess burden" costs of collecting money from taxpayers may be around 20 percent or more.

17. The number of years between invention and exploitation of certain inventions was, for example: television, twenty-two years; DDT, three; jet engine, fourteen; nylon, eleven; radar, thirteen; steam engine, eleven; and the fluorescent lamp, seventy-nine. Cited in "Returns to Research and Development Spending," a report by the U.K. Office of Science and Technology, May 1993.

18. Zvi Griliches, "Productivity, R&D, and the Data Constraint," *American Economic Review* (March 1994), pp. 1–23.

19. Consider federal housing subsidies, which have consumed billions of dollars over decades. According to John C. Weicher, in *Housing: Federal Policies and Programs* (Washington, D.C.: American Enterprise Institute, 1980), pp. 5–11, building public housing was originally conceived as a cure for social problems of the slums but has contributed little to that goal. Subsidized housing benefits its occupants, but spillover benefits appear to be nil.

20. Frank R. Lichtenberg, "R&D Investment and International Productivity Differences," Working paper 4161 (Cambridge, Mass.: National Bureau of Economic Research, September 1992).

21. Linda R. Cohen and Roger G. Noll, *The Technology Pork Barrel* (Washington, D.C.: Brookings Institution, 1991).

22. Francis Narin, Kimberly S. Hamilton, and Dominic Olivastro, CHI Research, "The Increasing Linkage between U.S. Technology and Public Science" (draft report), March 17, 1997. In this study, the term *publicly funded science* denotes research funded by government and nonprofits: everything except industrial research. It also includes foreign research. The great majority of patents do not cite any research articles, but those that do lean heavily on academic research.

23. See, for example, the testimony of Erich Bloch and Bruce Smith, U.S. Congress, House of Representatives, Subcommittee on Basic Research of the Committee on Science, *Hearings on Restructuring the Federal Scientific Establishment*, September 7, 1995.

24. "Prime Formula Weds Number Theory and Quantum Physics," *Science*, vol. 274, December 20, 1996, pp. 2014–15.

25. See, for example, Richard R. Nelson and Paul M. Romer, "Science, Economic Growth, and Public Policy," *Challenge* (March-April 1996), pp. 9–21.

26. U.S. Department of Commerce, *Statistical Abstract of the United States 1995*, table 624.

27. Christopher Hill, George Mason University, unpublished manuscript, 1996.

Chapter 4: Benefits from Science

1. Of course, a substantial share of the economic growth was the consequence of technical change, as we saw in chapter 2.

2. J. Mokyr and R. Stein, "Science, Health and Household Technology: The Effect of the Pasteur Revolution on Consumer Demand," National Bureau of Economic Research manuscript, April 30, 1994.

3. Some complain that medical research has raised the price of health care. Rapid increases in spending on health do not, however, necessarily denote higher

prices. Most likely, the price per unit of health care, properly defined, has actually fallen as new drugs, diagnostics, and procedures have been introduced. If you needed a serious operation, would you rather have it performed in 1997 or in 1897—with the prices but also the techniques of 1897?

4. Thanks to economic analysis, we have learned much about the most efficient policies for dealing with these problems.

5. See Michael Fumento, *Polluted Science: The EPA's Campaign to Expand Clean Air Regulations* (Washington, D.C.: AEI Press, 1997).

6. See Dennis T. Avery and Alex Avery, "Farming to Sustain the Environment," Hudson Institute briefing paper no. 190, May 1996. The authors also point out that liberalized trade in farm products is required to meet the soaring demand for food.

7. One can well imagine the reaction of today's social scientists to being offered grants only for designing psychological warfare techniques.

8. One reviewer pointed out that the framework for the analysis of the payoff from scientific research—including this book—derives from more fundamental social science research about market distortions, Pigouvian taxes, and the like.

9. Thomas C. Melzer, "President's Message," *Review*, Federal Reserve Bank of St. Louis (May/June 1996), p. 1.

10. Milton Friedman and Anna Schwartz, *The Great Contraction 1929–1933* (Princeton, N.J.: Princeton University Press, 1965).

11. Economists might debate whether these benefits would be captured by a Solow-type measure. For present purposes, it is enough to say that a sizable part of such benefits would not be so measured.

12. Other forms of benefits can be mentioned. Behavioral sciences, for example, are contributing to better health care by improving the understanding of the behavioral aspects of people's health.

13. See Rita R. Colwell, "Global Climate and Infectious Disease: The Cholera Paradigm," *Science*, vol. 274, December 20, 1996, pp. 2025–31. This discussion also draws on an unpublished manuscript by Steven Payson.

14. Testimony of Agriculture Department Under Secretary I. Miley Gonzalez before the House Appropriations Subcommittee on Agriculture, September 10, 1997.

15. Francis Fukuyama, *Trust: The Social Virtues and the Creation of Prosperity* (New York: Free Press, 1995).

16. Richard R. Nelson and Nathan Rosenberg, "Technical Innovation and National Systems," in Richard R. Nelson (ed.), *National Innovation Systems: A Comparative Analysis* (New York: Oxford University Press, 1993), pp. 4–5.

17. "Mass [achussets]. Considered Vulnerable to Research Cuts," *Boston Globe*, March 1, 1996.

18. This is not to deny the very tangible benefits that have come from these fields. Consider, for example, the contributions of astronomy to the development of the Global Positioning System that uses satellites to pinpoint precisely locations anywhere on earth and is now part of a multibillion dollar industry. Without "very long baseline interferometry," radio telescopes fixing on distant cosmic radio sources, satellites could not be positioned accurately enough to make the system work right.

Chapter 5: Limited Role for Government

1. The economic literature on public goods is extraordinarily large and complex. I have tried to introduce only those concepts required to set up criteria by which to evaluate the costs of downsizing federal science funding. For a recent exposition and interpretation, see Richard R. Nelson, "Why Bush's *Science: The Endless Frontier* Has Hindered Development of an Effective Civilian Technology Policy," in Claude E. Barfield (ed.), *Science for the Twenty-first Century: The Bush Report Revisited* (Washington, D.C.: AEI Press, 1997).

2. This diagram is adapted from Loren Yager and Rachel Schmidt, *The Advanced Technology Program: A Case Study in Federal Technology Policy* (Washington, D.C.: AEI Press, 1997), p. 19.

3. *Economic Report of the President, 1995* (Washington, D.C.: GPO), pp. 162–63.

4. R. H. Coase, "The Lighthouse in Economics, *Journal of Law and Economics*, vol. 17, no. 2 (October 1974), pp. 357–76.

5. See chapter 7 for further discussion.

6. See Charles Stuart, "Welfare Costs per Dollar of Additional Tax Revenue in the United States," *American Economic Review*, vol. 74 (June 1984), pp. 352–62. Stuart gives a wide range of estimates based on various assumptions about tax rates and the elasticity of labor supply. His mid-range is higher than 20 percent, but the nation's marginal tax rates are now lower than when Stuart wrote, which lowers the excess burden.

7. Chapter 9 has more to say about "competitiveness" as a justification for federal support of R&D.

8. This justification would not apply to the federal support for the education of medical doctors that intend to go into private practice rather than perform research. Some would give high priority to scientific education, even over research funding. See Paul Romer, "Beyond Market Failure," in *AAAS Science and Technology Yearbook* (Washington, D.C.: American Association for the Advancement of Science, 1997), p. 159.

9. Charles Murray, *What It Means to Be a Libertarian* (New York: Broadway Books, 1997), p. 58.

10. According to a General Accounting Office audit, 63 percent of the ATP's grantees did not approach a private sector lender before going to the government for money. U.S. General Accounting Office, *Measuring Performance: The Advanced Technology Program and the Private Sector* (GAO/RCED 96-47) (Washington, D.C.: GAO, February 15, 1996).

11. Recall that the cost of extracting funds from the taxpayer is probably around 20 percent.

12. See "OSTP Gears Up for Change," *Science*, May 2, 1997, pp. 668–70.

13. William A. Niskanen, "R&D and Economic Growth—Cautionary Thoughts," in Barfield, ed., *Science for the Twenty-first Century*, pp. 81–95.

14. Terence Kealy, *The Economic Laws of Scientific Research* (New York: St. Martin's Press, 1996).

15. Murray, *What It Means to Be a Libertarian*.

CHAPTER 6: JUST ANOTHER DOWNTURN?

1. E-mail from Alan Hale, ahale@nmsu.edu, March 21, 1997.
2. See National Science Board, *Science and Engineering Indicators, 1996* (Washington, D.C.: National Science Foundation, 1996), chap. 3.
3. Other choices were "only some confidence" and "hardly any confidence at all," ibid., chap 7.
4. A less positive finding: 52 percent of adults agreed with the statement that "many scientists make up or falsify research results to advance their careers or make money." See ibid.
5. Ibid.
6. Third International Mathematics and Science Study, cited in Office of Science and Technology Policy, *Science and Technology Shaping the Twenty-first Century* (Report to Congress), 1997, p. 119.
7. Neal Lane, "A Devil's Paradox: Great Science, Greater Limitations," in *Science and Technology Policy Yearbook, 1996–97* (Washington, D.C.: American Association for the Advancement of Science, 1997), p. 127. The interior quotation is of Carl Sagan.
8. National Science Board, *Science and Engineering Indicators, 1993*, p. 489.
9. Michael Fumento, *Science under Siege: How the Environmental Misinformation Campaign Is Affecting Our Laws, Taxes, and Our Daily Lives* (New York: William Morrow and Co., 1993).
10. Marcia Angell, *Science on Trial* (New York: W.W. Norton & Co., 1996).
11. Daniel Sarewitz, *Frontiers of Illusion: Science, Technology, and the Politics of Progress* (Philadelphia: Temple University Press, 1996).
12. Paul R. Gross and Norman Levitt, *Higher Superstition: The Academic Left and Its Quarrels with Science* (Baltimore: Johns Hopkins University Press, 1994).
13. Paul R. Gross and Norman Levitt (eds.), *The Flight from Science and Reason* (Baltimore: Johns Hopkins University Press, 1997).
14. John Horgan, *The End of Science: Facing the Limits of Knowledge in the Twilight of the Scientific Age* (New York: Addison-Wesley, 1996), p.16.
15. Charles J. Sykes, *ProfScam: Professors and the Demise of Higher Education* (Washington, D.C.: Regnery Gateway, 1988).
16. Lynne V. Cheney, *Telling the Truth: Why Our Culture and Our Country Have Stopped Making Sense—and What We Can Do about It* (New York: Simon & Schuster, 1995).
17. Ibid., p. 152.
18. Angell, *Science on Trial*, p. 177.
19. Fumento, *Science under Siege*.
20. Horgan, *The End of Science*, p. 6. Italics in original.
21. A parallel can be drawn with the House and Senate Intelligence Committees, which (except in special circumstances) are totally removed from the public eye. Members of these vital committees invest large amounts of time serving the national interest, valuable time that has virtually no political payoff among the voters back home.

Chapter 7: The Federal Laboratories

1. National Academy of Sciences, Committee on Criteria for Federal Support of Research and Development, *Allocating Federal Funds for Science and Technology* (Washington, D.C.: National Academy Press, 1995), p. 61.
2. Federal Laboratory Review Panel, Report to the Presidential Science Advisor and the White House Science Council, May 1983.
3. Secretary of Energy Advisory Board, *Alternative Futures for the Department of Energy National Laboratories*, the report of the Task Force on Alternative Futures for the Department of Energy National Laboratories (February 1995). Hereafter referred to as the "Galvin report."
4. Daniel S. Greenberg, "Shortchanging Science," *Washington Post*, February 19, 1997.
5. Some scientists doubt that these simulations will be accurate.
6. *Economist*, March 16, 1996, pp. 82–83.
7. The 1998 Energy Department budget, however, calls for $876 million to build a new "virtual" atomic testing facility, the National Ignition Facility at Lawrence Livermore National Laboratory, plus another $185 million to design a facility to produce tritium, needed to make new atomic warheads.
8. Congressional Budget Office, *Reducing the Deficit: Spending and Revenue Options*, February 1995, p. 111.
9. Galvin Report, p. 7.
10. *Science and Government Report*, May 15, 1997.
11. Irwin M. Stelzer with Robert Patton, *The Department of Energy : An Agency That Cannot Be Reinvented* (Washington, D.C.: AEI Press, 1996).
12. CRADA means Cooperative Research and Development Agreement, under terms established by the Stevenson-Wydler Technology Innovation Act (1980) and the Federal Technology Transfer Act (1986).
13. Stelzer, *The Department of Energy*.
14. An emerging issue: Should foreign companies be allowed to buy research on leading-edge technologies?

Chapter 8: Federal Role in Academic Research

1. David C. Mowery and Nathan Rosenberg, "The U.S. Innovation System," in Richard Nelson (ed.), *National Innovation Systems: A Comparative Analysis* (New York: Oxford University Press, 1993), p. 36.
2. The terms *university* and *academic research*, as used in this book, should be understood to include colleges.
3. Edwin Mansfield, "Academic Research Underlying Industrial Innovations: Sources, Characteristics, and Financing," *Review of Economics and Statistics*, vol. 77, no. 1 (1995), pp. 55–65.
4. National Science Board, *Science and Engineering Indicators, 1996* (Washington, D.C.: National Science Foundation, 1996); chapter 5 covers academic research.
5. Francis Narin et al., "The Increasing Linkage between U.S. Technology and Public Science," CHI Research, Inc., March 17, 1997.

6. Edwin Mansfield, "Academic Research and Industrial Innovation: A Further Note," *Research Policy*, vol. 21 (1992), pp. 295–96.

7. Paul Romer, "Beyond Market Failure," in *Science and Technology Policy Yearbook 1996–97* (Washington, D.C.: American Association for the Advancement of Science, 1997), p. 159.

8. National Science Foundation, *National Patterns of R&D Resources: 1996* (Washington, D.C.: NSF, 1996), p. 5.

9. Johns Hopkins and Stanford both manage large federal laboratories.

10. Wesley Cohen, Richard Florida, and Richard W. Goe, *University-Industry Research Centers in the United States*, Carnegie Mellon University, July 1994.

11. As reprinted in Richard R. Nelson, *The Sources of Economic Growth* (Cambridge: Harvard University Press, 1996), p. 227.

12. AUTM data cited by Rebecca S. Eisenberg, "Public Research and Private Development: Patents and Technology Transfer in Government-Sponsored Research," *Virginia Law Review* (November 1996), p. 1713.

13. The issue was whether regular mammograms should be recommended for women who are between forty and forty-nine years of age.

14. Other examples of political pressure coming to the aid of bad science are plentiful. Daniel S. Greenberg (*Washington Post*, May 6, 1997) cites the early 1950s case where the secretary of commerce fired the head of the National Bureau of Standards over a phony battery-recharging invention. Also around that time the president of the University of Illinois was fired for his refusal to support "research" on a fake cure for cancer developed by a faculty member with good political connections to the Board of Trustees and the state legislature.

15. U.S. Department of Education, *Mini Digest of Education Statistics*, table 29, 1995 (Education Department web pages).

16. National Science Board, *Science and Engineering Indicators, 1996*, p. 204.

17. Max Planck, *Scientific Autobiography and Other Papers* (New York: Philosophical Library, 1949), pp. 33–34.

18. National Research Council, "Enhancing Organizational Performance," draft report, 1997; cited in *Government Executive*, June 1997, p. 6.

19. *Time* magazine's cover story for March 17, 1997, was entitled "How Colleges Are Gouging: A Special Investigation into Why Tuition Has Soared."

CHAPTER 9: INDUSTRIAL RESEARCH

1. These figures come from National Science Foundation, *National Patterns of R&D Resources: 1996* (Washington, D.C.: NSF, 1997).

2. For a review of recent reports on this subject, see Bruce Smith, "The Clinton Approach: A New Linkage between Technology and Economic Policy?" in *Science and Technology Policy Yearbook* (Washington, D.C.: American Association for the Advancement of Science, 1994), pp. 19–30.

3. The average annual deviation from trend in company-funded R&D is 2.9 percent, versus 4.2 percent for producers' durable equipment (1960–1996). The coefficient of correlation between the two series is 0.96.

4. The NSF survey that measures industrial research spending was changed several years ago to include more firms in the service industries. This change

has possibly exaggerated the reported increases since that time, although the National Science Foundation believes that the data have been adjusted properly. More generally, data on R&D have a number of problems that make it unwise to put too much confidence in the accuracy of year-to-year changes.

5. For example, see *The Competitive Strength of U.S. Industrial Science and Technology: Strategic Issues* (Washington, D.C.: National Science Board, NSB-92-138, August 1992).

6. See Kenneth M. Brown, *The R&D Tax Credit: An Evaluation of Its Effectiveness* (Washington, D.C.: Joint Economic Committee of the Congress of the United States, 1985).

7. The data on federal funding of defense-related research by industry are problematic. See National Science Foundation, *National Patterns of R&D*, p. 63.

8. Richard S. Rosenbloom and William J. Spencer, "The Transformation of Industrial Research," *Issues in Science and Technology* (Spring 1996), p. 68. According to more recent accounts, however, there is some recognition that decentralization has problems; we may see some reversal of this in the next few years.

9. Charles F. Larson, "R&D in Industry," in *AAAS Report XIX, Research and Development FY 1995* (Washington, D.C.: American Association for the Advancement of Science, 1994), p. 30.

10. National Science Foundation, *National Patterns of R&D*, table C-6.

11. Bronwyn Hall, "The Private and Social Returns to Research and Development," in Bruce R. L. Smith and Claude E. Barfield, *Technology, R&D, and the Economy* (Washington, D.C.: Brookings Institution and AEI Press, 1996), pp. 140–62.

12. Zvi Griliches, "Comment," in *Brookings Papers: Microeconomics 2*, 1993, p. 394.

13. Rebecca Henderson and Iain M. Cockburn, "The Determinants of Research Productivity in Ethical Drug Discovery," in Robert B. Helms (ed.), *Competitive Strategies in the Pharmaceutical Industry* (Washington, D.C.: AEI Press, 1996), p. 184.

14. Wesley Cohen, Richard Florida, and W. Richard Goe, "University Industry Research Centers in the United States," Carnegie Mellon University, 1994.

15. For a good summary of the many such programs, see National Science Board, *Science and Engineering Indicators, 1996* (Washington, D.C.: National Science Foundation, 1996), pp. 4-18–4-20.

16. National Science Foundation, *National Patterns of R&D: 1996*, table C-2.

17. *Economic Report of the President, 1994*, p. 204. The report also cites the need for promoting U.S. competitiveness as a justification for the whole range of federal R&D spending.

18. The Commerce Department has a number of project evaluations underway. The only published evaluations deal mainly with how the program is perceived by grant recipients (they like receiving grants), rather than with overall benefits set against costs.

19. Loren Yager and Rachel Schmidt, *Evaluating Federal Technology Initiatives: The Advanced Technology Program* (Washington, D.C.: AEI Press, 1997).

20. In a report produced for the Commerce Department, economist Adam

Jaffe lays out this dilemma, advising the ATP to "fund projects that have a high rate of return, and a low probability that ATP funds are displacing private funds." He does not say explicitly how this could be done but recommends incorporating various factors into project choice, with the implication that NIST is currently not doing this. Adam B. Jaffe, "Economic Analysis of Research Spillovers: Implications for the Advanced Technology Program," NIST GCR 97-708, December 1996.

21. U.S. General Accounting Office, *Performance Measurement: The Advanced Technology Program and Private Sector Funding,* GAO-RCED-96-47.

22. Marcia Angell, *Science on Trial* (New York: W.W. Norton and Co., 1993), p. 203.

23. DuPont had supplied Teflon to a small company that used it to manufacture jaw implants. When legal problems arose, the company declared bankruptcy. Aggrieved patients then sued DuPont instead, which had supplied five cents worth of Teflon for each implant. Ibid., p. 84.

24. James T. Rosenbaum, "Lessons from Litigation over Silicone Breast Implants: A Call for Activism by Scientists," *Science,* June 6, 1997, p. 1524.

25. The expected costs of litigation (including some high imputed cost for the chance of being driven out of business) become part of overall research costs, and they reduce the rate of return on R&D.

26. National Academy of Sciences, *Allocating Federal Funds for Science and Technology* (Washington, D.C.: National Academy Press, 1995), p. 21.

Chapter 10: International Dimensions

1. All the international R&D comparisons use a conversion factor for calculating the parity of purchasing power for prices and exchange rates. This is generally recognized as the best approach, but to the extent that it is biased one way or the other, it tends to exaggerate the U.S. share of R&D spending.

2. National Science Board, *Science and Engineering Indicators, 1996* (Washington, D.C.: National Science Foundation, 1996), p. 5-33.

3. Ibid., p. 5-40.

4. Ibid., p. 5-36.

5. Includes those with dual citizenship.

6. National Science Board, *Science and Engineering Indicators, 1996,* p. 6-19.

7. Ibid., p. 2-23.

8. National Academy of Sciences, Committee on Criteria for Federal Support of Research and Development, *Allocating Federal Funds for Science and Technology,* Washington, D.C., 1995.

9. Data were available for only the nine largest OECD countries. I have estimated what the entire membership spent. My estimates are in line with those of the Commerce Department.

10. Alwyn Young, "The Tyranny of Numbers: Confronting the Statistical Realities of the East Asian Growth Experience," NBER working paper no. 4680, March 1994. For a contrary view, see Michael Sarel, *Growth in East Asia: What We Can and Cannot Infer* (Washington, D.C.: International Monetary Fund, 1996).

11. National Science Board, *Science and Engineering Indicators, 1996,* p. 2-23; 45

percent of the science and engineering Ph.D.s are granted in Europe and 35 percent in North America.

12. Ibid., p. 2-28.

13. National Science Foundation, *Human Resources for Science and Technology: The Asian Region* (NSF 93-303), 1993. While the quality of their Ph.D. programs trails that of U.S. universities, the leading Asian countries are clearly building up a base of human capital to support the technological growth they plan.

14. RAND, "A Framework for Analyzing Commercial Power Centers in Emerging Market Economies," DRR 1573 (draft report), 1997.

15. Paul Kennedy, *The Rise and Fall of Great Powers* (New York: Random House, 1987).

16. For a discussion of the declinist argument, see Joshua Muravchik, *Exporting Democracy: Fulfilling America's Destiny* (Washington, D.C.: AEI Press, 1992), chap. 5.

17. *Technology and the Federal Government: National Goals for a New Era*. Committee on Science, Engineering and Public Policy of the National Academy of Sciences, 1993.

18. *NSF in a Changing World: The National Science Foundation's Strategic Plan*, NSF 95-24, 1995.

19. Examples drawn from personal communication with William Blanpied and James Edwards, National Science Foundation.

20. Examples from Dorothy Nelkin, *Selling Science: How the Press Covers Science and Technology* (New York: W.H. Freeman and Co., 1995), p. 33.

21. See Paul Krugman, *Pop Internationalism* (Cambridge: MIT Press, 1996).

22. Scientific leadership does not correlate well with economic performance. Japan's economic boom of 1960-1990 was accomplished without world leadership, while for much of the postwar period U.S. growth rates trailed those of many of its scientific inferiors.

Chapter 11: Price of Downsizing Science

1. National Academy of Sciences, *Allocating Federal Funds for Science and Technology* (Washington, D.C.: National Academy Press, 1995).

2. Here I refer to macroeconomic data. There are still opportunities for researching the economics of particular research projects.

Index

Abramowitz, Moses, 22
Academic issues and research. *See* Universities
Advanced Technology Program (ATP), 50, 102, 103-4, 131
Advisory Commission to Study the Consumer Price Index, 24
Aeronautical research and technology activities, 102
Agricultural Research Service, 102
Agriculture, 12, 37
Agriculture, Department of. *See* Department of Agriculture
Allocating Federal Funds to Science and Technology (NAS), 106, 129
Alternative Futures for the Department of Energy Laboratories (1995). *See* Galvin Report
American Association for the Advancement of Science, 84, 97-98, 125
Angell, Marcia, 60, 105
Antitrust regulations, 47, 101
Apollo program, 69
Army Research Institute, 90
Asia, 15, 74, 111-14, 127, 144n13

Association of University Technology Managers, 86
ATP. *See* Advanced Technology Program

Baby-boom generation, 6
Bayh-Dole Act of 1980, 100-101
Bloch, Erich, 72
Brookhaven National Laboratory, 73
Budget issues. *See also* Economy, U.S.; balanced budget amendment, 5; budget deficit, 6, 26; budget plans, 5; discretionary spending, 7-8, 15, 55, 58, 84; economic growth, 26; federal laboratories, 64, 67; federal science and technology (FS&T), 64, 129; FY 1998 budget, 5, 7-8, 103; legislation, 133n5; priorities, 51; research and development, 28-29, 97-98, 128-30; targeting budget cuts, 29, 33, 129
Bureaucracy, 67, 72-73, 78
Bureau of Economic Analysis, 24, 25-26

145

Bureau of Labor Statistics, 22
Bush, Vannevar, 3-4, 12-13, 18

Carter administration, 15
Cheney, Lynne, 59-60
Chicago Academy of Sciences, 57, 58
China, 74, 113, 114-15
Clean Coal Technology Program, 103
Clinton administration, 5, 8, 103
Coalitions. *See* Collaboration and coalitions
Coase, Ronald, 47
Cockburn, Iain M., 100
Cohen, Linda, 30
Cohen, Wesley, 84, 100
Cold war, 6-7, 71
Collaboration and coalitions, 82, 85, 100-101, 117, 122-24
Commercial Space Transportation Office, 103
Competition and competitiveness, 96, 120-21
Competitiveness Technology Transfer Act of 1989, 101
Comprehensive Test Ban Treaty, 71
Computers, 24, 70, 135n9
Congress, 72, 76, 86
Consumer price index (CPI), 24
Corporate welfare, 103
Costs, 47-48, 123
Council for Competitiveness, 121
CPI. *See* Consumer price index

Defense Advanced Research Projects Agency, 102
Defense spending. *See* Research and development
Denison, Edward, 22
Department of Agriculture, 12, 40, 83
Department of Commerce, 102. *See also* Advanced Technology Program
Department of Defense, 64, 69, 73-74, 83, 102, 129-30
Department of Energy, 47, 65, 70-78, 83, 102-3, 123, 140n7
Department of Health and Human Services (DHHS), 36
Department of Transportation, 103
DHHS. *See* Department of Health and Human Services
DuPont, 105, 143n23

Economic issues. *See also* Budget issues; accelerator effect, 17, 31; analysis and measurement of data, 22-25, 35, 42; competitiveness, 121; displacement, 49-50, 67, 104; employment, 42, 56, 70, 88; gross domestic product, 22-23, 26, 35; growth, 112; knowledge as capital, 31; living standards, 3, 21; market adjustments, 74; productivity, 22, 28, 36, 41, 70, 99-100; rates of return, 25-28, 29, 33, 48-49, 51, 82; scientific research and, 3-4, 12-13, 18-20, 21-33, 82, 93; transfer payments, 49, 104
Economic Laws of Scientific Research, The (Kealy), 52
Economy, Asian, 111
Economy, U.S. *See also* Budget issues; budget pressures, 5-8, 128; downsizing, 125-32; federal support of R&D and, 4, 25-29, 33; interest on the federal debt, 7; peace dividend, 6-7; per capita income, 22; technological change and, 21-22
Edison Electric Institute, 85

Education, 41-42, 58. *See also* Universities
Eisenhower, Dwight D., 18
El Nino/southern oscillation (ENSO), 40
Endless Frontier, The. See Science: The Endless Frontier
End of Science, The (Horgan), 61
Energy crisis of 1970s, 30
Energy supply research and development, 102-3
Entitlement programs, 6, 7
Environmental programs, 37-38, 40
Environmentalists, 60
Europe, 111, 124

Federal government. *See* Funding
Federal Laboratory Review Panel, 67
Federally Funded R&D Centers (FFRDCs), 64
Federal Reserve Board, 39
Federal Technology Transfer Act of 1986, 100-101
Feminism, 60
FFRDCs. *See* Federally funded R&D Centers
Florida, Richard, 84, 100
Fossil Energy Research and Development, 103
France, 109
Fukuyama, Francis, 41
Funding, alternative sources, 86-87; cyclicality in federal funding, 16-17, 107; federal, 4, 5, 8, 9-12, 13-16, 28-30, 44, 46-48, 52, 64, 106-7; growth and downsizing, 11-12, 28-33; justifications for, 1; matching grants, 52; private, 31-33, 45-46, 54, 63-64; public approval and, 43, 61-62; targeting, 48-52, 54; tax credits, 52

Galvin Report, 67, 72, 73, 74, 76, 78, 79
Galvin, William, 67, 72
GDP (Gross domestic product). *See* Economic issues
Genetic engineering, 57
Germany, 109, 113
Goe, W. Richard, 100
Goods, public, definition, 44, 53; funding for, 46-47, 53; privatization and, 68; university research and, 82
Goodstein, David, 15
Government Performance and Results Act of 1993 (GPRA), 130
Gramm-Rudman Act, 5
Greenberg, Daniel, 69
Griliches, Zvi, 25, 28
Gross domestic product (GDP). *See* Economic issues

Hale, Alan, 56, 88
Hall, Bronwyn, 99
Harvard University, 83
Harwell Laboratory, 69
Health and medical research, 36-7, 59, 83, 125, 136n3
Henderson, Rebecca, 100
Hill, Christopher, 32
Hong Kong, 113-14
Horgan, John, 61
Human genome project, 69

IMF. *See* International Monetary Fund
Independence Day (film), 60
India, 113, 115
Indonesia, 115
Industry. *See also* Research and development; Advanced Technology Program, 50; basic research, 8, 54, 98-99, 122; benefits from R&D, 93-94; corpo-

147

rate funding, 95-96, 100, 101, 104-5, 106-7; downsizing, 90, 92-95, 99, 100, 101-7, 126; federal funding, 93, 97-98, 106-7; funding of academic research, 84, 85-86, 91, 92, 100, 101-2; government role, 94-95; management, 94, 98; national innovation system, 41, 95, 100; patents, 95, 100; productivity, 99-100; rate of return, 100; research and development, 81-82, 92, 95-101; subsidies, 99, 102-4, 106, 130-31
Industry, pharmaceutical, 100, 105
Information, 122-24
Internal Revenue Service (IRS), 47-48
International Monetary Fund (IMF), 111
IRS. *See* Internal Revenue Service

Japan, 109, 111, 112, 113, 114-15
Johns Hopkins University, 83

Kealy, Terence, 52
Kennedy, Paul, 116
Keyworth, George, 67
Korea, South, 112-13, 115

Laboratories, federal, closing commission, 77-78, 79; Department of Energy, 70-78; downsizing, 64-70, 75-78, 126; efficiency and management of, 72-74, 75-76; funding for, 51, 64; missions of, 66-67, 71, 74-75, 78-79, 131; privatization of, 67-69, 73-74, 77; sales to private sector, 76
Land-grant colleges, 12
Lane, Neal, 58
Large Hadron Collider (LHC), 123

Lawrence Livermore National Laboratory, 70, 71, 140n7
Leadership, 117-20, 131, 144n22. *See also* United States
Lederman, Leon, 55-56
LHC. *See* Large Hadron Collider
Lichtenberg, Frank R., 30
Lincoln Labs (MIT), 66
Living standards. *See* Economic issues
Lofgren, Zoe, 72-73
Los Alamos, 65-66, 70, 71

Manhattan Project, 69
Mansfield, Edwin, 25, 30, 81
Manufacturing Extension Partnership, 102
Medical research. *See* Health and medical research
Medicare, 6, 49
Melzer, Thomas C., 38-39
Merrell Dow, 105
Middle East, 74-75
Military preparedness, 98
Mitre Corporation, 73-74
Morrill Act, 12
Morse, Samuel, 12
Mowery, David C., 80
Multinational corporations, 48-49
Murray, Charles, 53

Nakamura, Leonard, 24
Narin, Francis, 30
NASA. *See* National Aeronautics and Space Administration
National Academy of Sciences, 118
National Aeronautics and Space Administration (NASA), 16, 65, 67, 69, 83, 102
National Cooperative Research Act of 1984, 100-101
National Ignition Facility, 70, 71, 140n7

National Institutes of Health (NIH), 8, 59, 65, 83
National Research Council, 90
National Science Foundation (NSF), data and reports, 5, 9, 11, 26, 54, 99, 114; funding for, 8; social sciences and, 38; U.S. leadership and, 118
Nelson, Richard, 41, 85
Nigeria, 74-75
NIH. *See* National Institutes of Health
Niskanen, William, 52
Nobel prizes, 110
Noll, Roger, 30
NSF. *See* National Science Foundation
Nuclear power, 57
Nuclear waste, 71
Nuclear weapons, 71

Oil, 74-75
Organization for Economic Cooperation and Development (OECD), 109, 111

Packard, David, 67
Packard Report, 72, 74
Partnership for a New Generation of Vehicles, 102
Patents and intellectual property rights, enforcement of patent law, 53; European, 110; federally funded projects, 30; government-government relationships and, 122-23; industrial research and, 95, 100; project excludability and, 46-47; royalties, 86; university research and, 82, 86-87, 91
Peer review, 51, 90
Pennsylvania, University of, 83
Persian Gulf War, 86

Ph.D.s. *See* Universities
Planck, Max, 89
Political issues, bad science, 141n14; discretionary spending, 8; federal laboratories, 64-65, 71-72; financial support for science, 4, 5, 17, 53, 62; university research, 86, 87
Prime numbers, 31
Privatization, 67-69. *See also* Funding
Productivity. *See* Economic issues
ProfScam: Professors and the Demise of Higher Education (Sykes), 59
Programs, 46-52
Publications, 109-10

R&D. *See* Research and development
RAND Corporation, 114-15
Rate of return. *See* Economic issues
Reagan administration, 15
Research and development (R&D). *See also* Economic issues; Science; Universities; benefits and effects of, 27, 30-31, 48-52, 103; defense, 5, 6-7, 8, 13, 16-17, 44, 66, 97-98, 110; definitions and models of, 9, 11, 20-21; displacement, 49-50; downsizing, 17-18, 27-29, 54, 55-62, 76-77, 92-95, 125-32; economic growth and, 21-24; efficiency of, 50-51, 72-74; federal, 63-79; federal funding for, 4, 5, 8, 9-12, 13-16, 28-30, 64, 93, 97-98; federal versus corporate versus university, 29-30; industrial, 8, 11, 14, 29-30, 32; as an investment, 25-28; joint ventures and collaboration, 47; president's budget, 8; productivity and,

22; public support for, 17-18; spending for, 11-12, 13-15; U.S. share of world spending, 14-15
Research, basic, applied research and, 13; cyclicality of, 17; definitions of, 9, 11; economic benefits, 98; federal government and, 54; "free riding on," 54, 99, 125; funding for, 8, 44-46, 66, 99; industrial, 8, 54, 98-99, 122; scientific advancements and, 3-4, 51, 53-54; space station, 69; universities and, 11, 81
Riemann Bernhard, 31
Rise and Fall of the Great Powers, The (Kennedy), 116
Romer, Paul, 31
Roosevelt, Franklin D., 12
Rosenberg, Nathan, 41, 80, 85, 98
Rosenbloom, Richard, 98
Russia, 111

Sandia National Laboratory, 65, 71
Schmidt, Rachel, 104
Science. *See also* Research and development; attitudes toward, 57, 139*n*4; "big science," 69-70; budget, 133*n*3; contributions of, 137*n*18; ignorance of, 58; junk science, 60-61; noneconomic benefits of, 35-43; policy and leadership, 116-17; quality of life and, 34
Science on Trial (Angell), 60, 105
Science: The Endless Frontier (Bush), 3-4
Science under Siege (Fumento), 61
Scientists and the scientific community. *See also* Universities; attitudes toward, 57, 60-61; downsizing and, 55-56; funding cuts and, 31; historically greatest, 11
Singapore, 113-14
Smoking, 36-37
Snow, C. P., 59
Social sciences, 38-39
Social security, 49
Solow, Robert, 21-22
Soviet Union, 17
Space program, 13, 15, 69
Spencer, William, 98
Stanford University, 83
Stevenson-Wydler Technology Innovation Act of 1980, 100-101
Stokes, Donald, 9
Subsidies. *See also* Funding; competitiveness and, 48; federal support of science, 53
Superconducting supercollider, 69, 119
Sykes, Charles, 59

Taiwan, 113-14
Tax issues, 52, 95, 97, 107
Telling the Truth (Cheney), 59-60
Tort system, 105, 143*n*25
Trade policy and issues, 117, 121, 123
Trust: The Social Virtues and the Creation of Prosperity (Fukuyama), 41

United Kingdom, 109, 113
United States. *See also* Economy; U.S. decline in scientific activity, 4, 115-17; federal role in scientific activity, 4; leadership in scientific research, 108-15, 120-24, 126-32; spending on R&D, 109, 111, 113, 114-15
Universities. *See also* Education; Scientists and the scientific com-

munity; academic fraud, 17-18; Asian, 114; basic research in, 11, 81, 82; downsizing and, 83, 84-91, 120, 127; federal funding of, 33, 80-81, 83-91, 132; graduate students and Ph.D.s, 15, 56, 82, 88-89, 99, 110, 127, 130; industrial funding of, 84, 85-86, 91, 92, 100; national laboratories, 124; patents and, 82, 86-87, 91; quality of research, 89-90; return on academic research, 25, 29-30, 82; role of, 11, 80, 82-84; science and engineering programs, 59-60; subsidization, 49; university-industry research centers, 84

What It Means to Be a Libertarian (Murray), 53

Yager, Loren, 104
Young, Alwyn, 112

Zeta function, 31

About the Author

KENNETH M. BROWN is a visiting fellow at the American Enterprise Institute for Public Policy Research, on leave from the National Science Foundation, where he headed the division that produces data on the nation's scientific enterprise. Previously he served on the National Intelligence Council at CIA. He was deputy under secretary for economic affairs in the Commerce Department and assistant Republican staff director with the Joint Economic Committee of Congress.

Mr. Brown's articles have appeared in the *Wall Street Journal*, the *Journal of Commerce*, *Regulation*, and economics journals. He has contributed articles to AEI volumes and edited an AEI publication on the tax credit for research and development. He received his Ph.D. in economics from the Johns Hopkins University. He served on the faculties of Ohio State University, the University of Notre Dame, and the University of Nairobi under the sponsorship of the Rockefeller Foundation.

Board of Trustees

Wilson H. Taylor, *Chairman*
Chairman and CEO
CIGNA Corporation

Tully M. Friedman, *Treasurer*
Tully M. Friedman & Co., LLC

Joseph A. Cannon
Chairman and CEO
Geneva Steel Company

Dick Cheney
Chairman and CEO
Halliburton Company

Harlan Crow
Managing Partner
Crow Family Holdings

Christopher C. DeMuth
President
American Enterprise Institute

Steve Forbes
President and CEO
Forbes Inc.

Christopher B. Galvin
CEO
Motorola, Inc.

Harvey Golub
Chairman and CEO
American Express Company

Robert F. Greenhill
Chairman
Greenhill & Co., LLC

Roger Hertog
President and COO
Sanford C. Bernstein and Company

M. Douglas Ivester
Chairman and CEO
The Coca-Cola Company

Martin M. Koffel
Chairman and CEO
URS Corporation

Bruce Kovner
Chairman
Caxton Corporation

Kenneth L. Lay
Chairman and CEO
Enron Corp.

Marilyn Ware Lewis
Chairman
American Water Works Co., Inc.

Alex J. Mandl
Chairman and CEO
Teligent, LLC

Craig O. McCaw
Chairman and CEO
Eagle River, Inc.

Paul H. O'Neill
Chairman and CEO
Aluminum Company of America

John E. Pepper
Chairman and CEO
The Procter & Gamble Company

George R. Roberts
Kohlberg Kravis Roberts & Co.

John W. Rowe
President and CEO
New England Electric System

Edward B. Rust, Jr.
President and CEO
State Farm Insurance Companies

John W. Snow
Chairman, President, and CEO
CSX Corporation

William S. Stavropoulos
Chairman and CEO
The Dow Chemical Company

James Q. Wilson
James A. Collins Professor of
 Management Emeritus
University of California at Los
 Angeles

The American Enterprise Institute for Public Policy Research

Founded in 1943, AEI is a nonpartisan, nonprofit, research and educational organization based in Washington, D. C. The Institute sponsors research, conducts seminars and conferences, and publishes books and periodicals.

AEI's research is carried out under three major programs: Economic Policy Studies; Foreign Policy and Defense Studies; and Social and Political Studies. The resident scholars and fellows listed in these pages are part of a network that also includes ninety adjunct scholars at leading universities throughout the United States and in several foreign countries.

The views expressed in AEI publications are those of the authors and do not necessarily reflect the views of the staff, advisory panels, officers, or trustees.

Officers

Christopher C. DeMuth
President

David B. Gerson
Executive Vice President

John R. Bolton
Senior Vice President

Steven P. Berchem
Vice President

Council of Academic Advisers

James Q. Wilson, *Chairman*
James A. Collins Professor of
 Management Emeritus
University of California at Los Angeles

Gertrude Himmelfarb
Distinguished Professor of History
 Emeritus
City University of New York

Samuel P. Huntington
Eaton Professor of the Science of
 Government
Harvard University

D. Gale Johnson
Eliakim Hastings Moore
 Distinguished Service Professor of
 Economics Emeritus
University of Chicago

William M. Landes
Clifton R. Musser Professor of
 Economics
University of Chicago Law School

Sam Peltzman
Sears Roebuck Professor of
 Economics and Financial Services
University of Chicago Graduate
 School of Business

Nelson W. Polsby
Professor of Political Science
University of California at Berkeley

George L. Priest
John M. Olin Professor of Law and
 Economics
Yale Law School

Thomas Sowell
Senior Fellow
Hoover Institution
Stanford University

Murray L. Weidenbaum
Mallinckrodt Distinguished
 University Professor
Washington University

Richard J. Zeckhauser
Frank Ramsey Professor of Political
 Economy
Kennedy School of Government
Harvard University

Research Staff

Leon Aron
Resident Scholar

Claude E. Barfield
Resident Scholar; Director, Science
 and Technology Policy Studies

Cynthia A. Beltz
Research Fellow

Walter Berns
Resident Scholar

Douglas J. Besharov
Resident Scholar

Robert H. Bork
John M. Olin Scholar in Legal
 Studies

Karlyn Bowman
Resident Fellow

Kenneth Brown
Visiting Fellow

John E. Calfee
Resident Scholar

Lynne V. Cheney
Senior Fellow

Dinesh D'Souza
John M. Olin Research Fellow

Nicholas N. Eberstadt
Visiting Scholar

Mark Falcoff
Resident Scholar

Gerald R. Ford
Distinguished Fellow

Murray F. Foss
Visiting Scholar

Michael Fumento
Resident Fellow

Diana Furchtgott-Roth
Assistant to the President and
 Resident Fellow

Suzanne Garment
Resident Scholar

Jeffrey Gedmin
Research Fellow

James K. Glassman
DeWitt Wallace–Reader's Digest
 Fellow

Robert A. Goldwin
Resident Scholar

Mark Groombridge
Abramson Fellow; Associate
 Director, Asian Studies

Robert W. Hahn
Resident Scholar

Kevin Hassett
Resident Scholar

Robert B. Helms
Resident Scholar; Director, Health
 Policy Studies

R. Glenn Hubbard
Visiting Scholar

James D. Johnston
Resident Fellow

Jeane J. Kirkpatrick
Senior Fellow; Director, Foreign
 and Defense Policy Studies

Marvin H. Kosters
Resident Scholar; Director,
 Economic Policy Studies

Irving Kristol
John M. Olin Distinguished Fellow

Dana Lane
Director of Publications

Michael A. Ledeen
Freedom Scholar

James Lilley
Resident Fellow

Lawrence Lindsey
Arthur F. Burns Scholar in
 Economics

John H. Makin
Resident Scholar;
Director, Fiscal Policy
 Studies

Allan H. Meltzer
Visiting Scholar

Joshua Muravchik
Resident Scholar

Charles Murray
Bradley Fellow

Michael Novak
George F. Jewett Scholar
 in Religion, Philosophy,
 and Public Policy;
Director, Social and
 Political Studies

Norman J. Ornstein
Resident Scholar

Richard N. Perle
Resident Fellow

William Schneider
Resident Scholar

William Shew
Visiting Scholar

J. Gregory Sidak
F. K. Weyerhaeuser Fellow

Christina Hoff Sommers
W. H. Brady, Jr., Fellow

Herbert Stein
Senior Fellow

Irwin M. Stelzer
Resident Scholar; Director,
 Regulatory Policy Studies

Daniel Troy
Associate Scholar

Arthur Waldron
Director, Asian Studies

W. Allen Wallis
Resident Scholar

Ben J. Wattenberg
Senior Fellow

Carolyn L. Weaver
Resident Scholar;
 Director, Social Security
 and Pension Studies

Karl Zinsmeister
Resident Fellow;
 Editor, *The American Enterprise*

*This book was edited by
Dana Lane of the publications staff
of the American Enterprise Institute.
The index was prepared by Julia Petrakis.
The text was set in Palatino, a typeface
designed by the twentieth-century Swiss designer
Hermann Zapf. Jennifer Lesiak set the type,
and Edwards Brothers, Incorporated,
of Lillington, North Carolina,
printed and bound the book,
using permanent acid-free paper.*

The AEI Press is the publisher for the American Enterprise Institute for Public Policy Research, 1150 Seventeenth Street, N.W., Washington, D.C. 20036; *Christopher DeMuth*, publisher; *Dana Lane*, director; *Ann Petty*, editor; *Leigh Tripoli*, editor; *Cheryl Weissman*, editor; Alice Anne English, production manager.